# Technique Builders and Problem-Solving Experiments

**HOLT, RINEHART AND WINSTON**
*Harcourt Brace & Company*

**Austin** • New York • Orlando • Atlanta • San Francisco • Boston • Dallas • Toronto • London

## Lab Authors

**Dave Jaeger**
Chemistry Teacher
Will C. Wood High School
Vacaville, CA

**Suzanne Weisker**
Science Teacher and Department Chair
Will C. Wood High School
Vacaville, CA

## Safety and Disposal Reviewer

**Jay A. Young, Ph.D.**
Chemical Safety Consultant
Silver Spring, MD

## Teacher Reviewers

**Cheryl Epperson**
Chemistry Teacher
Flour Bluff High School
Corpus Christi, TX

**Marilyn Lawson**
Chemistry Teacher/Science Department Chair
Gregory Portland High School
Portland, TX

**Cover:** Photo by Sam Dudgeon
**Illustrations:** Progressive Information Technologies

Printed in the United States of America

ISBN 0-03-051932-2

14 15  022 05 04

# LAB MANUAL C CONTENTS

# Working in the World of a Chemist

### Meeting today's challenges

Even though you have already taken science classes with lab work, you will find the two types of laboratory experiments in this book organized differently from those you have done before. The first type of lab is called an *Exploration,* and it helps you gain skills in lab techniques that you will use to solve a real problem presented in the second type of lab, which is called an *Investigation.* The *Exploration* serves as a *Technique Builder,* and the *Investigation* is presented as an exercise in *Problem Solving.*

Both types of labs refer to you as an employee of a professional company, and your teacher has the role of supervisor. Lab situations are given for real-life circumstances to show how chemistry fits into the world outside of the classroom. This will give you valuable practice with real-world skills, such as creating a plan with available resources, creating a budget, and writing business letters.

As you work on these labs, you will better understand how the concepts you studied in the chapters are used by chemists to solve problems that affect life for everyone.

## Explorations

The *Explorations* provide step-by-step procedures for you to follow, encouraging you to make careful observations and interpretations as you progress through the lab session. There is a series of items and questions at the end of each one that are designed to help you make sense of your data and observations and relate them to the key chemistry concepts being studied.

### What you should do before an Exploration

Preparation will help you work safely and efficiently. The evening before a lab, be sure to do the following.

- **Read the lab procedure** to make sure you understand what you will do.
- **Read the safety information** that begins on page vii, as well as that provided in the lab procedure.
- **Write down any questions** you have in your lab notebook so that you can ask about them before the lab begins.
- **Prepare all necessary data tables** so that you will be able to concentrate on your work when you are in the lab.

### What you should do after an Exploration

Most teachers require a lab report as a way of making sure that you understood what you were doing. Your teacher will give you specific details about how to organize the lab report, but most lab reports will include the following. A sample lab report you can use as a model is shown in Figure A.

- **title** for the lab
- **summary paragraph(s)** describing the purpose and procedure
- **data tables and observations** that are organized and comprehensive

- **worked-out calculations** with proper units
- **answers** that are boxed, circled, or highlighted, for items in the *Analysis and Interpretation, Conclusions,* and *Extensions* sections

## Investigations

The *Investigations* may seem quite different to you because they do not provide step-by-step instructions. The *Investigations* require you to develop your own procedure to solve a problem presented to your company by a client. As a part of the research team doing work for the client, you must decide how much money to spend on the project and what equipment to use. Although this may seem very difficult, the *Investigations* contain a number of clues about what to do to be successful in *Problem Solving.*

### What you should do before an Investigation

Before you will be allowed to work on the lab, you must turn in a preliminary report. Usually, you must describe in detail the procedure you plan to use, provide complete data tables for the data and observations you will collect, and list exactly what equipment you will need and the costs. Only after your teacher, acting as your supervisor, approves your plans are you allowed to proceed. A sample preliminary report is shown in Figure B on page vi. Before you begin writing a preliminary report, follow these steps:

Tamara Kieklak
Fifth Period
March 31, 1994

#### Lab Report
Title Exploration 12A—Boiling Point Elevation and Molar Mass

Summary  If a solute is dissolved in water, it will change the boiling point of water. First, I determined the boiling point of pure water, and then measured the boiling point with 10.0 g of each of the three solutes. The solutes used were NaCl, sucrose, and $MgSO_4$. Then I tested an unknown substance in the same manner. I compared the results to see what the unknown substance was.

Data Table

| Solution | Boiling Point |
|----------|---------------|
| $H_2O$ | 100.0°C |
| NaCl | 103.5°C |
| Sucrose | 100.5°C |
| $MgSO_4$ | 101.8°C |
| Unknown | 102.0°C |

Analysis and Interpretation

1. $\Delta t_b$ for NaCl = 103.5°C − 100.0°C = 3.5°C
   $\Delta t_b$ for Sucrose = 100.3°C − 100.0°C = 0.3°C
   $\Delta t_b$ $MgSO_4$ = 101.8°C − 100.0°C = 1.8°C
   $\Delta t_b$ Unknown = 101.9°C − 100.0°C = 1.9°C

2. The data for unknown most closely matches that for $MgSO_4$.

3. $m = \dfrac{\Delta t_b}{K_b}$

**Figure A    Sample Lab Report**

## Figure B   Sample Preliminary Lab Report

Stacey Rivera
Third Period
March 22, 1994

<u>Title</u> Boiling-Point Elevation—Contamination

<u>Summary</u> A sample of powdered sugar contaminated by another substance will be analyzed to determine the identity of the contaminant. Because the contaminant is either glucose, salt, or calcium chloride, the procedure can be a simple boiling-point elevation test. As a control for the procedure, the boiling point of pure water is determined by heating it with a Bunsen burner and measuring the temperature at boiling with a thermometer. Then the boiling point with 10.0 g of each of the possible contaminants is measured. Once the data is obtained, the same test for the boiling-point elevation of the unknown substance can be used to identify the contaminant.

<u>Data Table</u>

| Solution | Boiling Point |
|----------|---------------|
| $H_2O$ | |
| Glucose | |
| Salt | |
| $CaCl_2$ | |
| Unknown | |

<u>Materials</u>

| | |
|---|---|
| Lab space/fume hood/utilities | $15,000 |
| Disposal fee for 40 g of solutes | $80,000 |
| Beaker tongs | $1,000 |
| $CaCl_2$, 10 g | $5,000 |
| Glucose, 10 g | $5,000 |
| NaCl, 10 g | $5,000 |
| 250 mL flask | $1,000 |
| 100 mL graduated cylinder | $1,000 |
| Balance | $5,000 |

- **Read the *Investigation* thoroughly,** searching for clues.
- **Jot down notes** in your lab notebook as you find clues.
- **Consider what you must measure or observe** to solve the problem.
- **Think about *Explorations*** you have done that used a similar technique or reaction.
- **Imagine working through a procedure,** keeping track of each step and what equipment you will need.
- **Carefully consider** whether your approach is the best, most efficient one.

### What you should do after an Investigation

After you finish, organize a report of your data as described in the memorandum contained in the *Investigation*. This report is usually in the form of a one- or two-page letter to the client. (Your teacher may have additional requirements for your report.) Carefully consider how to convey the information that the client needs to know. In some cases, a graph or diagram can communicate information better than words can. As a part of your report, you must include an invoice for the client that explains how much they owe and how much you charged for each part of the procedure. Remember to include the cost of your labor during the analysis.

## Tips for Success in the Lab

Whether you are performing an *Exploration* or an *Investigation,* you can do the following to help ensure success.

- **Read each lab twice** before you prepare the data tables.
- **Make sure you understand** everything that will happen in the lab.
- **Read the safety precautions** in the lab and on pages vii–x.
- **Prepare the data tables** before you enter the lab.
- **Record all data and observations immediately** in your lab notebook.
- **Use appropriate units** when recording data.
- **Keep your lab desk organized** and free of clutter.

## Lab Report Grades

Lab report requirements and grading criteria vary from teacher to teacher. A good lab report usually has these characteristics.

### Explorations

- No significant errors in technique occurred during the lab.
- Data and observations are written accurately, descriptively, and completely.
- There are no errors in calculations.
- Good reasoning and logic are shown.
- Students seem to understand what the lab was about.
- Items from the *Analysis and Interpretation* and *Conclusions* sections are answered clearly, concisely, and accurately, with correct units given.
- Any graphs are drawn and labeled accurately and neatly.

### *Your teacher may have additional requirements.*

### Investigations

- Preliminary report shows careful and thorough planning with good reasoning and logic.
- Proposed procedure is appropriate and safe.
- Proposed data tables explicitly detail all measurements that must be made to solve the problem.
- Proposed list of equipment includes everything necessary and nothing unnecessary.
- No significant errors in technique occurred during the lab.
- Report or letter is concise, clear, and well organized.
- Data and analysis calculations are accurate, with correct units and no errors in the calculations.
- Percent error for quantitative answers is less than 15%.
- Students seem to understand what the lab was about.
- Any graphs are drawn and labeled accurately and neatly.
- Costs for materials actually used are kept under the budgeted figure.

### *Your teacher may have additional requirements.*

# Safety in the Chemistry Laboratory

## Chemicals are not toys

Any chemical can be dangerous if it is misused. Always follow the instructions for the experiment. Pay close attention to the safety notes. Do not do anything differently unless told to do so by your teacher.

Chemicals, even water, can cause harm. The trick is to know how to use chemicals correctly so that they will not cause harm. If you follow the rules stated in the following pages, pay attention to your teacher's directions, and follow the cautions on chemical labels and the experiments, then you will be using chemicals correctly.

## These safety rules always apply in the lab

1. **Always wear a lab apron and safety goggles.**
   Even if you aren't working on an experiment at the time, laboratories contain chemicals that can damage your clothing. Keep the strings of your lab apron tied. More importantly, some chemicals can cause eye damage, even blindness. If your safety goggles are uncomfortable or get clouded up, you are not the first person to have these problems. Ask your teacher for help. Try lengthening the strap a bit, washing the goggles with soap and warm water, or using an anti-fog spray.

2. **No contact lenses are allowed in the lab.**
   Even while wearing safety goggles, chemicals could get between contact lenses and your eyes and cause irreparable eye damage. If your doctor requires that you wear contact lenses instead of glasses, then you should wear eye-cup safety goggles in the lab. Ask your doctor or your teacher how to use this very important and special eye protection.

3. **NEVER WORK ALONE IN THE LABORATORY.**
   You should always do lab work *only* under the supervision of your teacher.

4. **Wear the right clothing for lab work.**
   Necklaces, neckties, dangling jewelry, long hair, and loose clothing can cause you to knock things over or catch items on fire. Tuck in neckties or take them off. Do not wear a necklace or other dangling jewelry, including hanging earrings. It isn't necessary, but it might be a good idea to remove your wristwatch so that it is not damaged by a chemical splash.

   Pull back long hair, and tie it in place. Nylon and polyester fabrics burn and melt more readily than cotton, so wear cotton clothing if you can. It's best to wear fitted garments, but if your clothing is loose or baggy, tuck it in or tie it back so that it does not get in the way or catch on fire.

   Wear shoes that will protect your feet from chemical spills—no open-toed shoes or sandals and no shoes with woven leather straps. Shoes made of solid leather or a polymer are much better than shoes made of cloth. Also, wear pants, not shorts or skirts.

5. **Only books and notebooks needed for the experiment should be in the lab.**
   Do not bring other textbooks, purses, bookbags, backpacks, or other items into the lab; keep these things in your desk or locker.

6. **Read the entire experiment before entering the lab.**
   Memorize the safety precautions. Be familiar with the instructions for the experiment. Only materials and equipment authorized by your teacher should be used. When you do the lab work, follow the instructions and the safety precautions described in the directions for the experiment.

7. **Read chemical labels.**
   Follow the instructions and safety precautions stated on the labels.

8. **Walk with care in the lab.**
   Sometimes you will have to carry chemicals from the supply station to your lab station. Avoid bumping into other students and spilling the chemicals. Stay at your lab station at other times.

9. **Food, beverages, chewing gum, cosmetics, and smoking are NEVER allowed in the lab.**
   (You should already know this.)

10. **NEVER taste chemicals or touch them with your bare hands.**
Also, keep your hands away from your face and mouth while working, even if you are wearing gloves.

11. **Use a sparker to light a Bunsen burner.**
Do not use matches. Be sure that all gas valves are turned off and that all hot plates are turned off *and* unplugged when you leave the lab.

12. **Be careful with hot plates, Bunsen burners, and other heat sources.**
Keep your body and clothing away from flames. Do not touch a hot plate after it has just been turned off. It is probably hotter than you think. The same is true of glassware, crucibles, and other things after you remove them from a hot plate, drying oven, or the flame of a Bunsen burner.

13. **Do not use electrical equipment with frayed or twisted cords or wires.**

14. **Be sure your hands are dry before using electrical equipment.**
Before plugging an electrical cord into a socket, be sure the electrical equipment is turned off. When you are finished with it, turn it off. Before you leave the lab, unplug it, but be sure to turn it off FIRST.

15. **Do not let electrical cords dangle from work stations; dangling cords can cause tripping or electrical shocks.**
The area under and around electrical equipment should be dry; cords should not lie in puddles of spilled liquid.

16. **Know fire drill procedures and the locations of exits.**

17. **Know the location and operation of safety showers and eyewash stations.**

18. **If your clothes catch on fire, walk to the safety shower, stand under it, and turn it on.**

19. **If you get a chemical in your eyes, walk immediately to the eyewash station, turn it on, and lower your head so that your eyes are in the running water.**
Hold your eyelids open with your thumbs and fingers, and roll your eyeballs around. You have to flush your eyes continuously for at least 15 min. Call your teacher while you are doing this.

20. **If you have a spill on the floor or lab bench, call your teacher rather than trying to clean it up by yourself.**
Your teacher will tell you if it is OK for you to do the cleanup; if it is not, your teacher will know how the spill should be cleaned up safely.

21. **If you spill a chemical on your skin, wash it off under the sink faucet, and call your teacher.**
If you spill a solid chemical on your clothing, brush it off carefully so that you do not scatter it on anyone else, and call your teacher. If you get a liquid on your clothing, wash it off right away if you can get it under the sink faucet, and call your teacher. If the spill is on your pants or somewhere else that will not fit under the sink faucet, use the safety shower. Remove the pants or other affected clothing while under the shower, and call your teacher. (It may be temporarily embarrassing to remove your pants or other clothing in front of your class, but failing to flush that chemical off your skin could cause permanent damage.)

22. **The best way to prevent an accident is to stop it before it happens.**
If you have a close call, tell your teacher so that you and your teacher can find a way to prevent it from happening again. Otherwise, the next time, it could be a harmful accident instead of just a close call.

23. **All accidents should be reported to your teacher, no matter how minor.**
Also, if you get a headache, feel sick to your stomach, or feel dizzy, tell your teacher immediately.

24. **For all chemicals, take only what you need.**
On the other hand, if you do happen to take too much and have some left over, DO NOT put it back in the bottle. If somebody accidentally puts a chemical into the wrong bottle, the next person to use it will have a contaminated sample. Ask your teacher what to do with any leftover chemicals.

25. **NEVER take any chemicals out of the lab.**
(This is another one that you should already know. You probably know the remaining rules also, but read them anyway.)

26. **Horseplay and fooling around in the lab are very dangerous.**
NEVER be a clown in the laboratory.

27. **Keep your work area clean and tidy.**
After your work is done, clean your work area and all equipment.

28. **Always wash your hands with soap and water before you leave the lab.**

29. **Whether or not the lab instructions remind you, ALL of these rules APPLY ALL OF THE TIME.**

## Safety Symbols

To highlight specific types of precautions, the following symbols are used in the *Explorations* and *Investigations*. **Remember that no matter what safety symbols are used, all of the 29 safety rules described previously should be followed at all times.**

- Wear laboratory aprons in the laboratory. Keep the apron strings tied so that they do not dangle.
- Wear safety goggles in the laboratory at all times. Know how to use the eyewash station. (Rules **1, 17,** and **19** apply. Which rule says not to wear contact lenses? Do regular eyeglasses provide enough protection?)

- Never taste, eat, or swallow any chemicals in the laboratory. Do not eat or drink any food from laboratory containers. Beakers are not cups, and evaporating dishes are not bowls. (Rules **9** and **10** apply.)
- Never return unused chemicals to the original container. (Which rule applies?)
- Some chemicals are harmful to our environment. You can help protect the environment by following the instructions for proper disposal.
- It helps to label the beakers and test tubes containing chemicals. (This is not another new rule, just a good idea.)

- Never transfer substances by sucking on a pipette or straw; use a suction bulb. (This one is another new rule.)
- Never place glassware, containers of chemicals, or anything else near the edges of a lab bench or table. (This is another new rule.)

Here is a question: Are there any rules other than **9** and **10** that apply in the lab when you see this safety symbol?

- If a chemical gets on your skin or clothing or in your eyes, rinse it immediately, and alert your teacher. (Rules **19** and **21** apply.)
- If a chemical is spilled on the floor or lab bench, tell your teacher, but do not clean it up yourself unless your teacher says it is OK to do so. (Rule **20** applies.)

Here is another question: What rules other than **19, 20,** and **21** apply in the lab when you see this safety symbol?

- When heating a chemical in a test tube, always point the open end of the test tube away from yourself and other people. (This is another new rule.)

WARNING! From here to the end of this safety section, it is up to you to look at the list of rules and identify whether a specific rule applies or if the rule presented is a new rule.

- Tie back long hair, and confine loose clothing. (Rule **?** applies.)
- Never reach across an open flame. (Rule **?** applies.)
- Use proper procedures when lighting Bunsen burners. Turn off hot plates, Bunsen burners, and other heat sources when not in use. (Rule **?** applies.)
- Heat flasks and beakers on a ring stand with wire gauze between the glass and the flame. (Rule **?** applies.)

- Use tongs when heating containers. Never hold or touch containers with your hands while heating them. Always allow heated materials to cool before handling them. (Rule **?** applies.)

- Turn off gas valves when not in use. (Rule **?** applies.)

- Use flammable liquids only in small amounts. (Rule **?** applies.)

- When working with flammable liquids, be sure that no one else in the lab is using a lit Bunsen burner or plans to use one. Make sure there are no other heat sources present. (Rule **?** applies.)

What other rules should you follow in the lab when you see this safety symbol?

- Check the condition of glassware before and after using it. Inform your teacher of any broken, chipped, or cracked glassware because it should not be used. (Rule **?** applies.)

- Do not pick up broken glass with your bare hands. Place broken glass in a specially designated disposal container. (Rule **?** applies.)

- Never force glass tubing into rubber tubing, rubber stoppers, or wooden corks. To protect your hands, wear heavy cloth gloves or wrap toweling around the glass and the tubing, stopper, or cork, and gently push in the glass. (Rule **?** applies.)

- Do not inhale fumes directly. When instructed to smell a substance, use your hand, wave the fumes toward your nose, and inhale gently. (Some people say "waft the fumes.")

- Keep your hands away from your face and mouth. (Rule **?** applies.)

- Always wash your hands before leaving the laboratory. (Rule **?** applies.)

Finally, if you are wondering how to answer the questions that asked what additional rules apply to the safety symbols, here is the correct answer. **Any time you see any of the safety symbols you should remember that all 29 of the numbered laboratory rules always apply.**

# Conservation of Mass

## Objectives

*Demonstrate* proficiency in detecting a chemical reaction.

*Measure* masses of reactants and products with a laboratory balance.

*Design* chemical experiments.

*Relate* observations of a chemical reaction to the law of conservation of mass.

## Situation

Your company has received a contract to develop a series of lab exhibits to be used at a *Fun Science* display that will be traveling to shopping malls across the country. The first exhibit requires participants to observe a chemical reaction between two household substances, vinegar and baking soda, and then design a way to contain the gas given off by the reaction so that the mass can be measured to determine if it is conserved.

## Background

The law of conservation of mass states that matter is neither created nor destroyed during a chemical reaction and that the mass of a system should therefore remain constant during any chemical process. This means that during any reaction the sum of the masses of the products of a reaction must be the same as the sum of the masses of the reactants. For example, imagine using a spark to ignite a match in a glass jar which is so tightly sealed that no matter can enter or escape. As the match burns, flammable substances from the match and some oxygen from the air are transformed into water, carbon dioxide, and other substances. Measurements taken before and after the match burned would show that the total mass remains the same.

Although this concept may seem obvious, it was not always so. When early scientists first began exploring chemical changes, they frequently did not consider the effects of the air and other gases. Similarly, the equipment originally used often did not take into account the possibility that a gas could be a product or a reactant.

## Problem

You will need to perform the following steps for this reaction to develop the exhibit.
- Accurately measure the mass of reactants and products in a controlled chemical reaction in an open container.
- Design a method to perform the same chemical reaction in a closed container so that no products escape.
- Accurately measure the mass of reactants and products in a controlled chemical reaction in a closed container.
- Compare your results with those of other groups.
- Provide instructions for the exhibit.

# Safety

 Always wear safety goggles and a lab apron to protect your eyes and clothing. If you get a chemical in your eyes, immediately flush the chemical out at the eyewash station while calling to your teacher. Know the locations of the emergency lab shower and eyewash station and the procedure for using them.

 Do not touch any chemicals. If you get a chemical on your skin or clothing, wash the chemical off at the sink while calling to your teacher. Make sure you carefully read the labels and follow the directions on all containers of chemicals that you use. Do not taste any chemicals or items used in the laboratory. Never return leftovers to their original containers; take only small amounts to avoid wasting supplies.

 Always clean up the lab and all equipment after use, and dispose of substances according to proper disposal methods. Wash your hands thoroughly before you leave the lab after all lab work is finished.

# Materials

- Baking soda ($NaHCO_3$)
- Vinegar (acetic acid, $CH_3COOH$, solution)
- 400 mL beaker
- 100 mL graduated cylinder
- Balance
- Clear plastic cups (or 150 mL beakers), 2
- Large plastic bag
- Twist ties, 2
- Weighing papers, 2

# Preparation

1. **Organizing Data**

   Make a data table in your lab notebook with four columns and three rows. Label the boxes in the first row of the second, third, and fourth columns *Initial mass (g)*, *Final mass (g),* and *Change in mass (g)*. Label the boxes in the second and third rows of the first column *Part 1* and *Part 2*. Below the table, leave room for observations about the reaction in Parts 1 and 2.

# Technique

**Part 1**

2. Place a piece of weighing paper on the laboratory balance. Place about 4–5 g of baking soda on the weighing paper. When you have that much, carefully transfer the baking soda to a plastic cup or beaker. *Never put any chemical directly on the balance pan of a laboratory balance. Never return any excess chemical to its original container because the risk that you might accidentally put the wrong chemical into the container is too great.*

3. Using the 100 mL graduated cylinder, measure about 100 mL of vinegar. Remember to read the volume at the bottom of the meniscus, the curved surface that the water forms where it meets the air. Pour the vinegar into the second plastic cup (or beaker).

4. Place both cups on the balance pan of a laboratory balance, and determine the initial total mass of the system to the nearest 0.01 g. Record this mass in your data table. (If the balance pan is too small to hold both cups, measure the mass of one, and then the other. Add these masses to find the total mass.)

5. Take the cups off the balance. Carefully pour the vinegar into the cup that contains the baking soda. Pour gently, along the inside wall of the baking soda cup, so splashing is kept to a minimum. It may help the reaction to reach completion sooner if you gently swirl the cup to make sure the reactants are well-mixed. Record your observations in your lab notebook. *Add only a small amount of vinegar at a time to avoid a strong reaction that causes splattering and the loss of part of the reactants or products.*

6. When the reaction has finished, place both cups back on the balance to determine the final mass of the system to the nearest 0.01 g. Record the final mass in your data table. Calculate any change in mass.

*Part 2*

7. Examine the plastic bag and the twist ties. Develop a procedure that will test the law of conservation of mass more accurately than does the procedure in Part 1. When you think you know what to do, ask your teacher to approve your plan. (Hint: it will be easier to measure the mass if you rest the plastic bag in the 400 mL beaker. Do not use more than 50 mL of vinegar.)

8. After your teacher approves your plan, implement it using the same materials and quantities as in Part 1. Use the plastic bag and the twist ties instead of the cups from Part 1. Remember to stop and record your data in your data table after each step.

9. If you were successful in step **8** and your results reflect that mass was conserved, proceed. If not, find a lab group that was successful, and find out what they did and why. Test their procedure to determine whether the results of the other group are reproducible.

*Cleanup and Disposal*

10. Place each product or reactant in the separate disposal containers provided by your teacher. Clean up your equipment and lab station. Thoroughly wash your hands after completing the lab work and cleanup.

## Analysis and Interpretation

1. **Organizing Conclusions**
   Describe all evidence indicating that a chemical change occurred in this experiment.

## Conclusions

2. **Evaluating Methods**
   Was the law of conservation of mass really violated in Part 1 as it appeared? Explain your answer.

3. **Evaluating Methods**
   Explain why the results for Part 2 were different from those for Part 1. What is the system in Part 1? What is the system in Part 2? How do they differ?

4. **Organizing Ideas**
   How can this lab be used to prepare the *Fun Science* exhibits? Explain your proposal for packaging the contents, and write instructions for the participants who will visit the exhibit.

**5. Evaluating Methods**

Explain how the steps of the scientific method were used in this procedure.

**6. Applying Ideas**

You can tell from the title of this Exploration that one of the themes that applies to this procedure is the theme of conservation. Choosing from the four themes listed below pick another theme, and explain how it relates to what happened during the procedure.

*a.* Macroscopic observations and micromodels

*b.* System interactions

*c.* Equilibrium and change

*d.* Classification and trends

## Extensions

**1. Applying Conclusions**

When a log burns, the resulting ash obviously has less mass than the log did. Explain whether this loss of mass violates the law of conservation of mass.

**2. Research and Communications**

The law of conservation of mass was first stated by Antoine Lavoisier in 1789. Prepare a report about Lavoisier's work and the textbook that he published.

**3. Research and Communications**

Just as mass is conserved in chemical reactions, so is energy. Find out several ways in which energy is converted and transferred when an automobile is driven. Present your findings as a poster to share with your class.

**4. Research and Communications**

The reaction that takes place in this Exploration can be summarized as follows.

$$NaHCO_3 + CH_3COOH \longrightarrow Na^+ \; CH_3COO^- + H_2O + CO_2$$

| sodium bicarbonate | acetic acid (dissolved) | sodium acetate (dissolved) | water | carbon dioxide gas |

Look up each substance in a chemical handbook and prepare a fact sheet that includes information such as melting point, boiling point, and density.

**5. Research and Communications**

Using a chemical handbook or other reference, compile a list of the hazards associated with each of the reactants and products. Prepare a presentation for the class about whether or not these chemicals are too dangerous to be used in a shopping mall. Thoroughly explain your reasoning.

# Separation of Mixtures

## Objectives

*Identify* chemical and physical properties of substances.

*Relate* knowledge of properties to the task of separating mixed items.

*Identify* as many methods for separating the items as possible.

*Separate* the components of a mixture.

*Analyze* success of methods for purifying mixtures.

*Design and implement* your own procedure.

## Situation

You work for a company that has joined the adopt-a-school program. The Springfield Consolidated School District is holding a Science Olympics, and your company has volunteered to prepare challenge packages containing a mixture of sand, salt, iron filings, and poppy seeds. The participants in the Science Olympics will be challenged to recover each of the components of the mixture as separate items. Your supervisor has asked you to try the challenge so that you can develop guidelines for evaluating the contestants' procedures.

## Background

A mixture is a combination of two or more kinds of matter. The different kinds of matter in a mixture can be separated by physical means because each component in a mixture retains its own composition and properties. Some methods will work for some components, but not for others. Another issue to keep in mind is how easy it will be to reclaim the different components of a mixture. For example, although paper chromatography can indicate the number and colors of components in a mixture, it can be difficult to use as a method for recovering the components.

## Problem

In order to prepare guidelines for evaluating the task, you must do the following.

- Identify as many physical and chemical properties of the substances in the mixture as you can.
- Figure out how to use these properties to identify as many different methods of separation as you can.
- Evaluate the methods to determine which is the best.
- Perform this method, and record the time it takes.

# Safety

 Always wear safety goggles and a lab apron to protect your eyes and clothing. If you get a chemical in your eyes, immediately flush the chemical out at the eyewash station while calling to your teacher. Know the locations of the emergency lab shower and eyewash station and the procedure for using them.

 Do not touch any chemicals. If you get a chemical on your skin or clothing, wash the chemical off at the sink while calling to your teacher. Make sure you carefully read the labels and follow the directions on all containers of chemicals that you use. Do not taste any chemicals or items used in the laboratory. Never return leftovers to their original containers; take only small amounts to avoid wasting supplies.

 Always clean up the lab and all equipment after use, and dispose of substances according to proper disposal methods. Wash your hands thoroughly before you leave the lab after all lab work is finished.

# Preparation

### Organizing Methods

1. Before you begin, you will need to develop a plan for separating the components of the mixture. Start by trying to determine which properties of a component in the mixture are not shared by most of the other components. When you think you know what to do, **write down the entire plan in your lab notebook before you proceed.** Be sure to estimate how long each step will take so that you can plan your time in lab more effectively.

2. If any part of your procedure involves a long period of waiting, try to reach that point before the end of the class period. Write your name on the side of your petri dish, and use it as a holding bin for your setup. Ask your teacher where you can leave it until you can complete your procedure on the next lab day.

# Technique

3. Obtain a sample of the mixture, a petri dish, a microfunnel, and an 8-well microchemistry strip. Using the microfunnel, place a small part of your sample in the first well of the 8-well microchemistry strip so that you will be able to compare your separated components to the original mixture. Place a small piece of tape over the opening of the well so that the contents will not spill out.

4. Record the time when you begin your work in your lab notebook.

5. Using any or all of the items listed in the materials list, implement the procedure you wrote in your lab notebook to separate and recover all four components of the rest of the mixture. Make as many observations as possible at each step, and record exactly what you do in your lab notebook.

# Materials

- Distilled water
- Sample of mixture
- 8-well micro-chemistry strip
- Aluminum foil
- Cellophane
- Cotton balls
- Filter paper
- Forceps
- Glass funnel
- Magnets
- Microfunnel
- Paper clips
- Paper towels
- Petri dish
- Pipets
- Plastic forks
- Plastic spoons
- Plastic straws
- Rubber stoppers
- Tape
- Test-tube holder
- Test-tube rack
- Test tubes
- Tissue paper
- Wood splints

6. When you have separated the first component from the mixture, place a small amount of it in the third well of the 8-well microchemistry strip, leaving the second well empty between the mixture in the first well and the purified component in the third well. Place a small piece of tape over the opening of the well.

7. As each component is separated, place small amounts in the fourth through sixth wells of the 8-well microchemistry strip. Place a small piece of tape over the opening of each well after it is filled, so that the contents do not spill out.

8. When all of the components have been separated, ask your teacher to inspect the samples and approve your work. If it is approved, record how long it has taken in your lab notebook. Write this on a small piece of paper. Place the paper and the 8-well microchemistry strip in your petri dish, and put them in the location indicated by your teacher.

9. Examine the strips of other lab groups. Record observations about the purity of each lab group's samples of salt, sand, iron filings, and poppy seeds. Record observations about the amount of time each lab group took.

## Cleanup and Disposal

10. Your teacher should have placed several disposal containers at the front of the room. After the appropriate wells have been filled with small samples of the recovered components, place the remainder of each recovered component in the appropriate disposal containers. Clean up your equipment and lab station. Thoroughly wash your hands after completing the lab work and cleanup.

## Analysis and Interpretation

1. **Analyzing Results**
Write a paragraph summarizing your procedure. In your estimation, rate the recovery of each component on a scale of 1–10. Justify the estimations of your success based on both the purity of your components and the time it took for you to do the job.

## Conclusions

2. **Evaluating Methods**
What made you decide to do your procedural steps in the order that you did them? Would any order have worked?

3. **Analyzing Methods**
If you were able to do the lab over again, what specific things would you do differently?

4. **Applying Ideas**
Name any materials or tools that were not available that might have made the separation of the substances easier.

5. **Analyzing Conclusions**
For each of the four components, describe a specific physical property that enabled you to separate it from the rest of the mixture.

**6. Evaluating Methods**

Discuss the relationship you expect to find between the speed of the process and the purity of the components recovered. Which do you think is more important? Explain your answer.

**7. Evaluating Methods**

Create a plan for scoring the results of students participating in the Science Olympics. Apply this plan to the observations you made about the other lab groups' results, and provide a list of scores for your class.

## Extensions

**1. Applying Information**

What methods could you use to determine the purity of each of your recovered components?

**2. Applying Ideas**

How would you separate each of the following two-part mixtures?

*a.* lead filings and iron filings
*b.* sand and gravel
*c.* sand and finely ground plastic foam
*d.* salt and sugar
*e.* alcohol and water
*f.* nitrogen and oxygen

**3. Research and Communication**

Find out about methods used to produce large quantities of products from coal or oil. Prepare a chart or poster to explain the different stages of separation of these products.

**4. Research and Communication**

Investigate the ways in which drinking water is purified in your community. Prepare a presentation to your class about how these methods work and why they must be used.

HOLT
ChemFile
LAB PROGRAM

# Separation of Mixtures — Tanker Truck Spill

January 23, 1995

Marissa Bellinghausen
Director of Investigations
CheMystery Labs, Inc.
52 Fulton Street
Springfield, VA 22150

Dear Ms. Bellinghausen:

A tanker truck carrying approximately $3.8 \times 10^4$ L of waste from a food-processing plant overturned along an embankment at the edge of Zavala Lake. The tanker carried no manifest, but its contents were apparently ground coffee, garlic powder, and mineral oil. Although these spilled products are not believed harmful, they make the water look and smell unpleasant. The lake is a favorite recreational spot used by many people for swimming, sailing, and water-skiing, so this is a problem that needs to be cleaned up.

Preliminary sample analysis of the lake water suggests that sand and/or charcoal may be adequate for cleaning the water. We are requesting that your company analyze the sample to evaluate the effectiveness of sand and charcoal as treatment mediums and to suggest detailed procedures for cleanup. A sample of the contaminated water will arrive under separate cover, as will a contract for $70 000.

Sincerely,

*Anita Camacho*

Anita Camacho
Director
Department of Health Services
Verde County

# Memorandum

**Date:** May 10, 1995
**To:** Angelo Portelli
**From:** Marissa Bellinghausen

*CheMystery Labs, Inc.*
*52 Fulton Street*
*Springfield, VA 22150*

If we do a good job planning the purification, we will probably be <u>hired to help implement it</u>, which will provide <u>steady revenue</u> for the company. This is <u>crucial</u> to helping us stay in business.

I think you should start by letting the sample <u>settle</u>. Then, I recommend some <u>filtering</u> steps. Because the final cleanup will involve huge volumes of water, I don't think using a filtering system with filter paper will be practical. Instead, consider a system in which water is pumped through several tubes of material. To model this system, pour the water into a <u>long-stemmed microfunnel</u> with about 15 g of <u>sand</u> and another long-stemmed microfunnel with about 8 g of <u>charcoal</u>. (Hint: to keep these substances in the microfunnel, place a wad of tissue paper or cotton in the bottom first.)

Because we <u>don't have a large budget</u>, I'll need to look over your plan before you proceed. Prepare a proposal that includes the following.
- detailed one-page <u>plan for the procedure</u>
- illustrations or descriptions of your filter bed constructions
- <u>listing of the properties</u> of the mixture's components that you will use to separate them
- list of all <u>necessary materials and equipment</u>, along with individual and total <u>costs</u> for the accounting department

When you are finished with your analysis, prepare a report to Anita Camacho in the form of a <u>two-page</u> <u>letter</u> that includes the following.
- <u>summary of your plan</u> to clean up the lake that re<u>lates the properties</u> of the components to your plan and reflects the <u>best use of filters</u>.
- justification of your <u>proposed filtering setup</u>
- discussion of the <u>benefits of sand</u> as a water purifier and the <u>benefits of charcoal</u> as a purifier
- identify the <u>sand filter construction that mini</u>mizes filtration time while maximizing water purity and explain why it works best
- data on <u>color, clarity, odor</u>, presence of <u>oils or solids, and volume</u> for each step of your procedure
- <u>percentage</u> of initial volume of $H_2O$ remaining after treatment
- discussion of the <u>likelihood that impurities re</u>main in the water
- <u>detailed invoice</u> for all costs

## Required Precautions

 Always wear safety goggles and a lab apron to protect your eyes and clothing.

 Do not touch any chemicals. If you get a chemical on your skin or clothing, wash the chemical off at the sink while calling to your teacher.

 Always wash your hands thoroughly when finished.

## Spill Procedures/ Waste Disposal Methods

- To recover the sand, hold the sand-filter funnel upside down over the sand disposal container, squeeze the tip, and roll it between your fingers. Use a micro-tip pipet to squirt water through the opening to flush the sand into the container.
- To recover the charcoal, pinch and roll the tip of the pipet, and then use a small piece of wire to push any that remains into the charcoal disposal container.
  - Oily residue on equipment should be absorbed onto paper towels, which may be placed in the trash can. Then wash and scrub the equipment with soap and water.
  - Purified water may be poured down the sink.
  - Any foul water, oil, or other liquids should be collected in the same disposal container.

## MATERIALS for Verde County
### REQUIRED ITEMS
( You must include all of these in your budget.)

| | |
|---|---|
| Lab space/fume hood/utilities | 15 000/day |
| Standard liquid disposal fee | 20 000/L of product |

### REAGENTS AND ADDITIONAL EQUIPMENT
(Include in your budget only what you will need.)

| | |
|---|---|
| Charcoal | 500/g |
| Sand (mixture of coarse and fine) | 500/g |
| 25 mL graduated cylinder | 1 000 |
| 50 mL beaker | 1 000 |
| 250 mL beaker | 1 000 |
| Desiccator | 3 000 |
| Glass stirring rod | 1 000 |
| Long-stemmed microfunnel | 1 000 |
| Pipet bulb without stem | 1 000 |
| Pipet (thin-stemmed) | 1 000 |
| Plastic spoon | 500 |
| Plastic tray | 1 000 |
| Ring stand/ring | 2 000 |
| Scissors | 500 |
| Six test tubes/holder/rack | 2 000 |
| Stopwatch | 5 000 |
| Strainer | 1 000 |
| Tissue paper | 500/piece |
| Wash bottle | 500 |
| Watch glass | 1 000 |
| Wire | 500 |
| Wooden splint | 500 |

\* No refunds on returned chemicals or unused equipment.

### FINES

| | |
|---|---|
| OSHA safety violation | 2 000/incident |

## References

Reread the theory behind mixtures and their separation discussed in your textbook. The physical properties of materials can be used to separate mixtures . To separate mixtures that form layers, squeeze a pipet bulb that has most of its stem cut off to pull the mixture into the bulb. Hold the bulb with the open side down, allow the layers to settle, then gently squeeze to empty each layer, one at a time, into different containers. Coarse sand filters out fewer particles and allows water to flow through faster than fine sand. To separate coarse sand from fine sand, pass it through a strainer held over a plastic tray. To measure volume, calibrate a wooden splint so that each mark on the stick corresponds to the height of 1.0 mL in the test tubes you will be using.

HOLT
ChemFile
LAB PROGRAM

# Covalent and Ionic Bonding— Ceramics Fixative

February 17, 1995

Mr. Reginald Brown
Director of Materials Testing
CheMystery Labs, Inc.
52 Fulton Street
Springfield, VA 22150

Dear Mr. Brown,

In our research for creating a new type of glaze for our ceramic pottery, we have run into a problem. We have tried using a variety of compounds to develop a fixative for the glaze but have failed to find one that meets our criteria.

The compound in the fixative must be transparent when it dissolves and white when it dries so that it will not affect the other colors in the glaze. Next, the compound must have a high melting point for the kiln firing. We would like the fixative to dissolve in water but not in alcohol. Finally, the compound must be electrically conductive when it is dissolved in water. It does not matter whether the fixative is organic or inorganic. We have identified four compounds for study: sucrose, sodium chloride, sodium carbonate, and salicylic acid. You should determine which of these substances are ionic and which are covalent and recommend which will best fit our criteria.

As discussed on the telephone, a contract for $135 000 will arrive under separate cover.

Sincerely,

*Kathleen Sylva*

Kathleen Sylva
Head Researcher
Ceramic Artisans

# Memorandum

*Date:* February 21, 1995
*To:* Teresa Rodriguez
*From:* Reginald Brown

CheMystery Labs, Inc.
52 Fulton Street
Springfield, VA 22150

This company is on the <u>cutting edge</u> of new artistic ceramics materials. If we succeed in doing well on this contract, <u>others could follow</u>.

Because we do not have a kiln or the necessary equipment to do high-temperature studies, I recommend that you <u>gently heat</u> about 0.5 g of each substance <u>in a test tube</u> held in a burner flame. <u>Once the substance melts, remove it</u> from the burner flame. If a substance does not appear to melt after 1.0 min, remove it from the flame. Although we will not be able to measure precise melting points, this will help us determine which <u>substances have the highest melting points</u>.

In order to help ensure quality results, let me review the following <u>before you begin work</u>.
- paragraph <u>summarizing your procedure</u>, including <u>safety precautions</u> you will take
- <u>data table</u> with spaces for all necessary items
- complete list of all <u>materials and equipment</u>, with total and individual <u>costs</u>

When you complete your research, prepare a summary of your results in the form of a <u>two-page letter</u> to Ms. Sylva. Be sure to include the <u>following</u>.
- <u>your recommendation</u> of the best compound
- paragraph discussing your <u>reasons</u> for your recommendation
- paragraph <u>summarizing the procedures</u> used, including a discussion of <u>how</u> the <u>criteria</u> requested <u>were tested</u>
- completed <u>data table</u>
- <u>detailed invoice</u> for materials and services

# Required Precautions

 Always wear safety goggles and a lab apron to protect your eyes and clothing.

 Tie back loose clothing and hair. Do not heat glassware that is broken, chipped, or cracked. Be sure test tubes are heat-resistant before heating. Always use a test-tube holder when heating a tube. Continuously move the tube while heating it to prevent pressure from building up. Never point the opening of a test tube that you are heating toward yourself or others.

 Ethanol is a flammable liquid. Perform the melting point tests first. Then, only after your teacher tells you that all students have finished their melting point tests and all Bunsen burners are turned off and have cooled down, proceed with testing how well the substances dissolve in water and ethanol.

 Do not touch any chemicals. If you get a chemical on your skin or clothing, wash the chemical off at the sink while calling to your teacher.

 Always wash your hands thoroughly when finished.

## References

Reread the portions of your textbook that discuss ionic and covalent bonds, and their properties.

The solubility of 1.0 g of a substance can be determined by shaking it with 5 mL of water or alcohol in a stoppered test tube.

Be sure to rinse the conductivity tester with distilled water between each test.

Use the total mass of the substances you test, including those dissolved in solvents, when calculating disposal costs.

## Spill Procedures/ Waste Disposal Methods

- Contact your teacher if a spill occurs.
- Each undissolved sample should go in its own separate disposal container.
- All water solutions should be placed in their own separate container.
- All ethanol solutions should be placed in their own separate container. Do not pour ethanol solutions down the drain.
- Scrub all of the test tubes well with soap, water, and a test-tube brush to remove any residues of the compounds tested.

## MATERIALS for Ceramic Artisans

### REQUIRED ITEMS
(You must include all of these in your budget.)

| | |
|---|---|
| Lab space/fume hood/utilities | 15 000/day |
| Standard solid disposal fee | 2 000/g |
| Balance | 5 000 |
| Bunsen burner/related equipment | 5 000 |

### REAGENTS AND ADDITIONAL EQUIPMENT
(Include in your budget only what you will need.)

| | |
|---|---|
| Ethanol | 1000/mL |
| Salicylic acid | 1 000/g |
| Sodium carbonate | 1 000/g |
| Sodium chloride | 1 000/g |
| Sucrose | 1 000/g |
| 150 mL beaker | 1 000 |
| 250 mL beaker | 1 000 |
| 250 mL flask | 1 000 |
| 100 mL graduated cylinder | 1 000 |
| Conductivity probe, LEAP | 5 000 |
| Conductivity tester, battery-operated | 5 000 |
| Forceps | 500 |
| Glass stirring rod | 1 000 |
| Rubber stopper (for test tube) | 500 |
| Six test tubes/holder/rack | 2 000 |
| Spatula | 500 |
| Wash bottle | 500 |
| Weighing paper | 500/piece |

\* No refunds for returned chemicals or unused equipment.

### FINES

| | |
|---|---|
| OSHA safety violation | 2 000/incident |

# Viscosity of Liquids

## Objectives

*Demonstrate* proficiency in comparing the viscosity of various liquids under identical test conditions.

*Construct* a small viscosimeter.

*Measure* flow time of various single-weight oils.

*Measure* the mass and volume of the oils to calculate density.

*Calculate* the relative viscosity of the oils.

*Graph* experimental data.

*Compare* viscosities and densities to determine the SAE rating of each oil.

## Situation

You have been contacted by an automotive service shop that received a shipment of bulk containers of motor oil. The containers had been shipped by freight train, but several boxcars had leaky roofs. As a result, the labels peeled off the cans. Before the shop uses this oil in cars, they must match up the cans with the types of oil that were listed on the shipping invoice, based on the viscosity and the SAE rating of the oils.

## Background

Viscosity is the measurement of a liquid's resistance to flow. Several factors contribute to viscosity. Liquids with high intermolecular forces tend to be very viscous. For example, glycerol has a high viscosity because of its tendency to form many hydrogen bonds. For other molecules, such as oils, the longer the chain length of the molecule, the more viscous they are. The longer chains not only provide greater surface area for intermolecular attractions, but also they can be intertwined more easily. For example, gasoline, which contains molecules that are chains of three to eight carbon atoms, is much less viscous than grease, which usually contains molecules with about 20 to 25 carbon atoms.

The Society of Automotive Engineers rates lubricating oils according to their comparative viscosities. These numerical values, called SAE ratings, range from SAE-10 (low viscosity) to SAE-60 (high viscosity) for oils typically used in combustion engines such as those in automobiles and trucks. The ratings are achieved with an instrument called a viscosimeter, which has a small capillary tube opening. The amount of time for a specific amount of motor oil to flow through the opening is a measure of viscosity. The less viscous oils flow through in a shorter time than the more viscous oils do.

## Problem

To match the correct oil sample to its SAE rating, you will need to do the following.

- Make your own viscosimeter from a pipet.
- Measure the relative viscosities of several oils by timing the oil as it flows through your viscosimeter.
- Measure mass and volume of each oil to calculate density.
- From the measurements, infer which labels belong on the containers of oil.

## Safety

Always wear safety goggles and a lab apron to protect your eyes and clothing. If you get a chemical in your eyes, immediately flush the chemical out at the eyewash station while calling to your teacher. Know the locations of the emergency lab shower and eyewash station and the procedure for using them.

Do not touch any chemicals. If you get a chemical on your skin or clothing, wash the chemical off at the sink while calling to your teacher. Make sure you carefully read the labels and follow the directions on all containers of chemicals that you use. Do not taste any chemicals or items used in the laboratory. Never return leftovers to their original containers; take only small amounts to avoid wasting supplies.

When you use a Bunsen burner, confine any long hair and loose clothing. Do not heat glassware that is broken, chipped, or cracked. Use tongs or a test-tube holder to handle heated glassware and other equipment because hot glassware does not always look hot. If your clothing catches on fire, WALK to the emergency lab shower, and use it to put out the fire. Because the oil tested in this lab is flammable, it should never be heated directly over a flame. Instead, use a hot-water bath, and never heat it above 60°C.

Pins are sharp; use them with care to avoid cutting yourself or others.

Always clean up the lab and all equipment after use, and dispose of substances according to proper disposal methods. Wash your hands thoroughly before you leave the lab after all lab work is finished.

## Preparation

1. **Organizing Data**

   Prepare a data table in your lab notebook with 13 columns and 8 rows. Label the boxes in the first row *Sample, Beaker mass (g), Total mass (g), Volume (mL), Trial 1—cool (s), Trial 2—cool (s), Trial 3—cool (s), Trial 1—room temp. (s), Trial 2—room temp. (s), Trial 3—room temp. (s), Trial 1—warm (s), Trial 2—warm (s),* and *Trial 3—warm (s).* In the first column, label the second through eighth boxes *A, B, C, D, E, F,* and $H_2O$. Leave three spaces below the table to record the cool temperature, the room temperature, and the warm temperature.

2. With a wax pencil, label each 50 mL beaker-test tube-pipet set with the name of one oil sample (*A, B, C, D, E,* or *F*). Label an additional set $H_2O$.

3. Place two marks 2.0 cm apart on the side of the bulb of the pipet, as shown in Figure A. The top mark will be the starting point and the lower mark will be the endpoint.

## Materials

- Distilled water
- Ice
- Oil samples, 10 mL, 6
- 50 mL beakers, 7
- 400 mL beakers, 2
- 10 mL graduated cylinder
- Pin, straight
- Pipets, thin-stem, 7
- Ruler, metric
- Stopwatch or clock with second hand
- Test-tube holder
- Test-tube rack
- Test tubes, small, 7
- Wax pencil

**Bunsen burner option**

- Bunsen burner and related equipment
- Ring stand and ring
- Wire gauze with ceramic center

**Hot plate option**

- Hot plate

**Thermometer option**

- Thermometer, non-mercury
- Thermometer clamp

**Probe option**

- Thermistor probe

**Chem**File

Pin hole in top of pipet bulb

Marks, 2 cm apart

Thin-stem pipet

**Figure A**

**4.** Carefully make a small hole in the top of the bulb of each pipet with the pin, as shown in Figure A. Be sure the hole is well above the marks you made on the side of the pipet bulb. Make the hole the same size for each pipet by putting in the pin the same way for each one. You will control the flow of oil with your finger and this hole.

## Technique

**5.** Measure the masses of all seven 50 mL beakers. Record them in your data table.

**6.** Pour about 5.0 mL of distilled water into the graduated cylinder. Measure and record the volume to the nearest 0.1 mL, and pour it into the $H_2O$ beaker.

**7.** Measure and record the mass of the $H_2O$ beaker with water in your data table.

**8.** Squeeze the $H_2O$ pipet bulb and fill the pipet with distilled water to above the top line. After it is full, place your finger over the pin hole. Place the pipet over the $H_2O$ beaker, lift your finger off the hole, and allow the liquid to flow into the beaker until the meniscus is even with the top line on the pipet bulb. Cover the hole promptly when the water reaches this point. Several practice trials may be necessary.

**9.** One member of the lab group should hold the pipet with a finger over the pinhole, and the other should use a clock with a second hand or a stopwatch to record precise time intervals. Hold the pipet over the $H_2O$ beaker. When the timer is ready, remove your finger from the pinhole, and allow the liquid to flow into the beaker until it reaches the bottom line on the pipet bulb. Record the time elapsed to the nearest 0.1 s in your data table in the section for room temperature. (If you do not have a stopwatch, measure the time elapsed to the nearest 0.5 s.) It may take several practice trials to master the technique.

**10.** Repeat steps **6–9** with each oil, using the appropriately labeled pipets and beakers. You should perform several trials for each oil and for water to obtain consistent results.

**11.** Using one of the 400 mL beakers, make an ice bath. Fill the test tubes to within 1.0 cm of the top with the appropriate oil or distilled water. Cool the samples for 5–8 min so that they are at a temperature between 0°C and 10°C. The key is that all of the samples must be at the same temperature. Measure the temperature of the water sample to the nearest 0.1°C with a thermometer or a thermistor probe and record it below your data table.

**12.** Repeat steps **6–9** with each of the cooled samples. Be sure to use the pipets and 50 mL beakers designated for each oil or distilled water. Record the volume, mass, and time elapsed for each trial in your data table.

13. Using a Bunsen burner or a hot plate and the second 400 mL beaker, prepare a warm-water bath with a temperature between 35°C and 45°C. If you measure the temperature with a thermometer, use a thermometer clamp attached to a ring stand to hold the thermometer in the water.

14. Refill the test tubes to within 1.0 cm of the top with the appropriate oil or distilled water. Place these test tubes into the warm-water bath, and allow the oil and water to warm. Record the temperature of the water sample with a thermometer or a thermistor probe when you remove the samples.

15. Repeat steps **6–9** with the warm samples. Record the volume, mass, and time elapsed for each trial in your data table.

## Cleanup and Disposal

16. Your instructor will have set out twelve disposal containers; six for the six types of oil and six for the pipets. **Do not pour oil down the sink. Do not put the oil or oily pipets in the trash can.** The distilled water may be poured down the sink. The test tubes should be washed with a mild detergent and rinsed. Always wash your hands thoroughly after cleaning up the area and equipment.

## Analysis and Interpretation

1. *Organizing Data*
   Determine the density of each sample.

2. *Organizing Data*
   Find the average flow time for each sample at each temperature.

3. *Analyzing Information*
   Calculate the relative viscosity of your samples at room temperature by applying the following formula. The values for the absolute viscosity of water are in units of centipoises (cp). A centipoise is equal to 0.01 g/cm·s.

$$\text{relative viscosity}_{oil} = \frac{\text{density}_{oil} \times \text{time elapsed}_{oil} \times \text{viscosity}_{H_2O}}{\text{density}_{H_2O} \times \text{time elapsed}_{H_2O}}$$

| Temperature (°C) | Absolute Viscosity for $H_2O$ (cp) |
|---|---|
| 18 | 1.053 |
| 20 | 1.002 |
| 22 | 0.955 |
| 24 | 0.911 |
| 25 | 0.890 |
| 26 | 0.870 |
| 28 | 0.833 |

# Conclusions

**4. Inferring Conclusions**

According to the invoice, the service station was supposed to receive equal amounts of SAE-10, SAE-20, SAE-30, SAE-40, SAE-50, and SAE-60 oil. Given that the oils with the lower SAE ratings have lower relative viscosities, infer which oil samples correspond to the SAE ratings indicated.

**5. Organizing Information**

Prepare a graph with flow time at room temperature on the $y$-axis and SAE rating on the $x$-axis.

**6. Organizing Information**

Prepare a graph with density on the $y$-axis and SAE rating on the $x$-axis.

**7. Organizing Information**

Prepare a graph with viscosity at room temperature on the $y$-axis and SAE rating on the $x$-axis.

**8. Organizing Information**

Prepare a graph with viscosity at room temperature on the $y$-axis and density on the $x$-axis.

**9. Inferring Conclusions**

How does temperature affect the viscosity of each sample?

**10. Interpreting Graphics**

Is there a relationship between density and viscosity?

**11. Interpreting Graphics**

What is the relationship between SAE rating and viscosity?

**12. Interpreting Graphics**

What is the relationship between viscosity and flow time?

# Extensions

**1. Predicting Outcomes**

Estimate what flow times you would measure at each temperature if you repeated the tests in this lab with SAE-35 oil.

**2. Relating Ideas**

Malcolm is trying to get the last of the pancake syrup out of a bottle. What can he do to make the syrup come out of the bottle faster? Explain how your plan will take advantage of viscosity.

**3. Research and Communication**

Contact a manufacturer of lubrication products such as Valvoline or Pennzoil, and write a short paper on the development and properties of the oils used in this investigation.

HOLT
ChemFile
LAB PROGRAM

# Viscosity—New Lubricants

February 21, 1995

Ms. Sandra Fernandez
Director of Development
CheMystery Labs, Inc.
52 Fulton Street
Springfield, VA 22150

Dear Ms. Fernandez:

Our company manufactures a wide variety of industrial lubricants. We produce lubricating oils used in land vehicles, aircraft, and ships. Our range of products includes hydraulic fluids, engine oils, and lubricating grease.

We are presently interested in developing two lubricants for use in an experimental, high-rpm engine to be used in a pump. We want your firm to research and develop two multigrade oils, which are mixtures of other oils. One of these mixtures should have a relative viscosity of 20 cp at approximately 5°C. The second mixture should have a relative viscosity of 10 cp at approximately 21°C.

Once you test and identify the best method for mixing the oils to achieve the desired properties, we are prepared to manufacture and market these lubricants on a large scale. In exchange for your work, we are willing to pay up to $150 000.

Sincerely,

*Judy Showalter*

Judy Showalter
Director of Research
General Lubricants

# Memorandum

**CheMystery Labs, Inc.**
*52 Fulton Street*
*Springfield, VA 22150*

**Date:** February 22, 1995
**To:** John X. Muhammad
**From:** Sandra Fernandez

We are still paying off several business loans and can certainly use this contract from General Lubricants. It will be your team's task to determine the ideal formulation of oils to achieve the two desired viscosities. Because of the number of research teams working on this project and the limited supply of oils available, you are limited to 50 mL of any single grade. Remember to work carefully to produce quality results.

Before you begin the job, I need to review the following items from you.

- detailed one-page plan for the procedure and all necessary data tables
- four proposed formulations for the new oils (Hint: you may wish to refer to data from Exploration 3-2 as you propose formulations.)
- detailed list of the equipment and materials you will need, along with the individual and total costs

Once your plan is approved, you may begin your lab work. Make at least 20 mL of each of the oil mixtures. When completed, prepare a report in the form of a two-page letter to Ms. Showalter that includes the following items.

- suggestion for multigrade oil to achieve the desired viscosities
- detailed and organized analysis of your test oil: density of samples, ratios of each grade used in samples, flow time, approximate SAE rating, and the average relative viscosity for each sample
- summary paragraph describing how you solved this problem
- detailed and organized data table
- detailed invoice including costs for equipment, supplies, and services

## Required Precautions

 Always wear safety goggles and a lab apron to protect your eyes and clothing.

 Do not touch any chemicals. If you get a chemical on your skin or clothing, wash the chemical off at the sink while calling to your teacher.

 When you use a Bunsen burner, confine any long hair and loose clothing. Do not heat glassware that is broken, chipped, or cracked. Use tongs or a test-tube holder to handle heated glassware and other equipment. Because the oil tested in this lab is flammable, it should never be heated directly over a flame. Instead, use a hot-water bath, and never heat it above 60°C.

 Pins are sharp; use them with care to avoid cutting yourself or others.

## Spill Procedures/ Waste Disposal Methods

- In case of spills, follow your teacher's directions. Any oil spills should be wiped up with paper towels.
- Any unmixed oil sample should be placed in its own designated container.
- Put samples of mixed oil in the disposal containers designated by your teacher.
- Do not pour any oil samples down the drain or place them in the trash can.

## References

Reread your textbook for information on intermolecular forces in nonpolar molecules, such as those found in oils. The viscosity testing procedure your team recently completed will be useful here. The data you gathered at that time can be used to help you select a promising mix of oils that should have properties close to those specified in the letter. You may need to calculate some relative viscosities for the pure oils using the Exploration's data for temperatures other than room temperature. The viscosity of water at 5°C is 1.52 cp, and at 21°C, it is 0.98 cp. Calculations of density are discussed in your textbook.

## MATERIALS for General Lubricants

### REQUIRED ITEMS
*(You must include all of these in your budget.)*

| | |
|---|---|
| Lab space/fume hood/utilities | 15 000/day |
| Standard liquid disposal fee | 20 000/L |

### REAGENTS AND ADDITIONAL EQUIPMENT
*(Include in your budget only what you will need.)*

| | |
|---|---|
| Ice | 500 |
| Oil samples (SAE-10, -20, -30, -40, -50, and -60) | 1 000/mL |
| 50 mL beaker | 1 000 |
| 400 mL beaker | 1 000 |
| 25 mL graduated cylinder | 1 000 |
| Aluminum foil | 1 000/cm² |
| Bunsen burner/related equipment | 10 000 |
| Duct tape | 500/cm |
| Filter paper | 500/piece |
| Glass stirring rod | 1 000 |
| Hot plate | 8 000 |
| Pin | 500 |
| Pipet | 1 000 |
| Ring stand/ring/wire gauze | 2 000 |
| Ruler | 500 |
| Six test tubes/holder/rack | 2 000 |
| Stopwatch | 5 000 |
| String | 500/cm |
| Thermistor probe | 2 000 |
| Thermometer clamp | 1 000 |
| Thermometer, nonmercury | 2 000 |
| Wax pencil | 500 |
| Weighing paper | 500/piece |

\* No refunds on returned chemicals or unused equipment.

### FINES

| | |
|---|---|
| OSHA safety violation | 2 000/incident |

# Chemical Reactions and Solid Fuel

## Objectives

***Recognize*** several different reaction types.

***Write*** balanced chemical equations for reactions observed in the laboratory.

***Prepare*** a gel, and determine whether it is homogeneous or heterogeneous.

***Calculate*** the mass of a fuel before burning and of the residue remaining after burning.

***Evaluate*** the lab-produced solid fuel by measuring how well it heats water compared to a commercially available fuel.

## Situation

An entrepreneur, Ms. Toietta LaFaye, has contracted your company to research a new camping/emergency fuel. The idea came from an incident in which several campers forgot to pack the fuel for their camp stove. Because the area prohibited the use of campfires, the campers needed to use an alternative fuel source. One of the campers made a gel that they used as solid fuel. Some chalk was crushed and mixed with vinegar. This mixture was filtered through a napkin, and the filtrate was heated using a solar reflector. Some rubbing alcohol was poured into the solution to form a gel, which burned.

Ms. LaFaye wants to provide campers with a ready-made fuel. She wants you to prepare the fuel under controlled conditions in the laboratory to determine how and why it works, compare the product to a commercially available solid fuel, and identify what other products are produced as the fuel is made and used.

## Background

One way to understand this process is to break it down into individual chemical reactions, as shown below. Ingredients are identified with placeholder names that can be replaced after they have been identified. Chalk is made mostly of calcium carbonate, $CaCO_3$. One of the main ingredients of vinegar is acetic acid, $CH_3COOH$. Rubbing alcohol is isopropanol, $CH_3CHOHCH_3$. It could be the flammable isopropanol in the gel that causes it to burn.

$$CaCO_3(s) + CH_3COOH(aq) \longrightarrow \text{Solution of Compound V} + \text{??}$$

$$\text{Compound V}(aq) + CH_3CHOHCH_3(l) \longrightarrow \text{Gel W} + \text{??}$$

$$\text{Gel W burns in air} \longrightarrow \text{Ash X}(s) + \text{??}$$

Ms. LaFaye also wants you to research whether ethanol, $CH_3CH_2OH$, would work better than isopropanol in the mixture.

## Problem

In order to provide Ms. LaFaye with a complete report that gives her all of the information she needs, you must do the following.

• Prepare two kinds of gel in the lab: one with ethanol and one with isopropanol, and measure the mass of the gel produced.

• Compare this gel to commercially available fuels by measuring how hot they make similar amounts of water.

• Measure the mass of ash remaining after complete burning.

• Use what you know about chemical reactions from your textbook to classify and describe each reaction.

## Materials

- Chalk, 25 g
- Commercially available solid fuel
- Distilled water
- Ethanol, $CH_3CH_2OH$, 30 mL
- Isopropanol, $CH_3CHOHCH_3$, 30 mL
- Vinegar (acetic acid solution), 100 mL
- 250 mL beakers, 6
- 100 mL graduated cylinder
- Balance
- Crucible tongs
- Evaporating dishes, 2
- Glass stirring rod
- Mortar and pestle
- Ring stands with ring clamps, 3
- Wire gauze with ceramic center, 3

**Gravity filtration option**

- Glass funnel
- Filter paper

**Vacuum filtration option**

- Aspirator for spigot
- Büchner funnel
- Filter paper
- One-hole rubber stopper or sleeve
- Vacuum flask and tubing

**Bunsen burner option**

- Bunsen burner and related equipment

**Hot plate option**

- Hot plate

**Probe option**

- Thermistor probes, 3

**Thermometer option**

- Thermometers, non-mercury, 3

## Safety

 **Always wear safety goggles and a lab apron to protect your eyes and clothing. If you get a chemical in your eyes, immediately flush the chemical out at the eyewash station while calling to your teacher. Know the locations of the emergency lab shower and eyewash station and the procedure for using them.**

 **Do not touch any chemicals. If you get a chemical on your skin or clothing, wash the chemical off at the sink while calling to your teacher. Make sure you carefully read the labels and follow the directions on all containers of chemicals that you use. Do not taste any chemicals or items used in the laboratory. Never return leftovers to their original containers; take only small amounts to avoid wasting supplies.**

 **When you use a Bunsen burner, confine any long hair and loose clothing. Do not heat glassware or other equipment that is broken, chipped, or cracked. Use tongs or a hot mitt to handle heated glassware and other equipment because hot glassware does not always look hot. If your clothing catches on fire, WALK to the emergency lab shower, and use it to put out the fire.**

 **Both ethanol and isopropanol are flammable. While working with ethanol and isopropanol to make the gelled or solid fuels, make sure there are no flames or other sources of ignition anywhere in the room. You may not obtain or work with ethanol or isopropanol until after all groups have finished heating the chalk-vinegar mixture and have turned off their burners or hot plates and allowed them to cool. No tests of the solid fuels should be done until all lab groups are finished making their fuels, all excess ethanol and isopropanol is stored in closed containers in the hood, the hood window is closed, and the hood fan is on.**

 **Always clean up the lab and all equipment after use, and dispose of substances according to proper disposal methods. Wash your hands thoroughly before you leave the lab after all lab work is finished.**

## Preparation

**1. Organizing Data**
Prepare two data tables in your lab notebook. The first data table should have three columns and four rows. Label the boxes in the first row *Measurement, Ethanol fuel,* and *Isopropanol fuel.* In the first column, label the second through fourth rows *Mass of empty evaporating dish (g), Mass of gel + dish (g),* and *Mass of ash + dish (g).*

**2. Organizing Data**

The second data table should have four columns and seventeen rows. Label the boxes in the first row *Time (min)*, *Temp. (°C)—ethanol fuel*, *Temp. (°C )—isopropanol fuel*, and *Temp. (°C)—commercial solid fuel*. Label the second through seventeenth boxes in the first column with the numbers 0–15. You will also need room below the data table to record your observations about the synthesis of the fuel and how it burned.

**3.** Measure the masses of the evaporating dishes to the nearest 0.01 g. Designate one to be used for the gelled isopropanol fuel and the other for the gelled ethanol fuel. Handle these dishes only with crucible tongs, even when cool, so that you will not cause errors by adding oil or dirt to them.

## Technique

**4.** Place about 25 g of chalk in a mortar and pestle. Using a gentle circular motion, grind it into small pieces.

**5.** Completely transfer the chalk to a 250 mL beaker. Add about 100 mL of vinegar to the beaker, and stir for 5 to 8 min. Record your observations below your data table.

**6.** Set up your filtering apparatus. If you are using a Büchner funnel for vacuum filtration, follow step **6a.** If you are using a glass funnel for gravity filtration, follow step **6b.**

**a. Vacuum filtration**

Screw an aspirator nozzle into the faucet. Attach a piece of thick plastic tubing to the side arm of this nozzle. Attach the other end of the thick plastic tubing to the side arm of the filter flask. Place a one-hole rubber stopper on the stem of the funnel, and fit the stopper snugly in the neck of the filter flask, as shown in Figure A. Place the filter paper on the bottom of the funnel, so that it is flat and covers all of the holes.

**Figure A**

### b. Gravity filtration

Use a glass funnel with filter paper as shown in Figure C, supported by a ring stand and ring. Prepare the filter paper by folding it as shown in Figure B. Fit the filter paper in the funnel, and wet it with a little water. Gently but firmly press the paper against the sides of the funnel so that no air can get between the funnel and the paper. The filter paper should not extend above the edge of the funnel.

**Figure B**

**Figure C**

7. Filter the chalk-vinegar mixture. If you are using a Büchner funnel for vacuum filtration follow step **7a.** If you are using a glass funnel for gravity filtration, follow step **7b.**

### a. Vacuum filtration

Prepare the filtering apparatus by pouring approximately 100 mL of water through the filter paper. After the water has gone through the funnel, empty the filter flask into the sink. Reconnect the filter flask, and pour the chalk-vinegar mixture into the funnel, using a stirring rod, as shown in the illustration.

### b. Gravity filtration

Place a 250 mL beaker at the bottom of the glass funnel. Prepare the filtering apparatus by pouring approximately 100 mL of water through the filter paper. Pour the water that is in the beaker after passing through the filter into the sink. Place the beaker back at the bottom of the glass funnel so that it will collect the filtrate from the funnel. Pour the chalk-vinegar mixture into the funnel, using a stirring rod, as shown in the illustration.

8. Pour about 40 mL of the clear filtrate into the ethanol fuel evaporating dish. If you will be performing the lab during two days, reserve another 40 mL of the filtrate for later. If you will be performing the lab all in one day, pour another 40 mL of the clear filtrate into the isopropanol fuel evaporating dish. Remember to use crucible tongs to hold and carry the evaporating dishes. Do not use your hands.

9. Gently heat the contents of the evaporating dish or dishes for a few minutes with a hot plate or a Bunsen burner until the volume is reduced by half. Turn off the hot plate or burner when finished, and allow it to cool. **Do not proceed with the next step until your teacher tells you to continue.**

10. Add 30 mL of ethanol to the ethanol fuel evaporating dish. The gel should form in about 8 s. Decant any excess ethanol into the ethanol disposal container. Measure the mass of the dish and gel. Record this and any observations about the gel and its synthesis in your data table. If you will be performing the lab during two days, proceed with step **13** today to test the ethanol-based fuel. Then, continue with step **11** when you are ready to make the isopropanol fuel.

11. Repeat step **10** using isopropanol to make the gelled fuel in the isopropanol fuel evaporating dish. Decant any excess isopropanol into the isopropanol disposal container. Measure the mass of the dish and gel. Record this and any observations about the gel and its synthesis in your data table.

12. **Do not proceed with the next step until all students have finished making their fuel, the disposal containers have been placed in the fume hood, and your teacher tells you to continue.**

13. Set up a ring stand and ring clamp for each fuel sample to be tested. Adjust each clamp so that its height is about 10.0 cm from the base of the ring stand. Place a wire gauze with ceramic center on each ring clamp. Set a 250 mL beaker on top of each gauze. Under one beaker, place the ethanol fuel. Under another, place the isopropanol fuel. Under the third, place the commercially available solid fuel.

14. Using a graduated cylinder, measure 75.0 mL of distilled water into each of the beakers above the fuel samples. (The volume does not have to be exactly 75.0 mL, but the volume of water in each beaker must be the same.)

15. Using three thermometer clamps and three thermometers or thermistor probes, suspend a thermometer or thermistor probe in each beaker so that it does not touch the sides or bottom of the beaker. Record the initial temperature of all of the water samples. If they are not all the same, wait a few minutes until they are.

16. When you are ready to continue, ask your teacher to light the fuels for you. Record the temperature of the water in each beaker at 1.0 min intervals for 15 min or until the fuels burnout.

17. Wait until the fuels have stopped burning and the evaporating dishes have cooled thoroughly, and then measure the mass of the ash and record it in your data table.

## Cleanup and Disposal

18. Place the ash in its own disposal container. Any excess alcohol should already be in the ethanol disposal container or the iso-propanol disposal container. The filter paper and the solid material on it may be placed in the trash, and any excess filtrate from the vinegar-chalk mixture may be poured down the drain. Always wash your hands thoroughly after cleaning up the area and equipment.

## Analysis and Interpretation

1. **Analyzing Observations**
   Explain what observations indicated that a chemical reaction was taking place when you added the vinegar to the chalk.

2. **Analyzing Information**
   Look at the first equation in the Background section at the beginning of this Exploration. Write a balanced equation for this reaction, including formulas for the products. (Hint: the reaction is a double-displacement reaction. When acetic acid is dissolved in water, some of it dissociates into acetate ions and hydrogen ions.)

3. **Analyzing Information**
   One of the products from the equation you wrote in item **2** underwent a decomposition reaction. One of the products was water. Write the balanced chemical equation for this decomposition reaction. How does this equation relate to your observations in item **1?**

4. **Analyzing Observations**
   Was there any evidence that the heating of the filtrate caused a chemical reaction?

5. **Analyzing Observations**
   Was there any evidence that the addition of the alcohols to form a gel (the second equation listed in the Background section) caused a chemical reaction?

6. **Analyzing Observations**
   Is the gel a homogeneous or heterogeneous mixture? Explain your answer.

7. **Analyzing Information**
   Assuming that the part of the gels that actually burned was the alcohol, write balanced equations for each combustion reaction.

8. **Analyzing Observations**
   Is the ash that remains after burning your gels represented as a product in the chemical equations written in item **7?** What is the primary component of the ash?

## 9. Organizing Data
How many grams of the component described in item **8** were contained in each gel sample? Calculate the corresponding percentages by mass.

## 10. Organizing Data
How many grams of each alcohol were present in the gel samples? How many moles of each alcohol were present?

## 11. Predicting Outcomes
The amount of energy produced by the combustion of exactly one mole of ethanol has been measured to be about $-1367$ kJ. The combustion of one mole of isopropanol releases about $-2006$ kJ. Calculate how much energy should have been produced as the gelled fuels burned. According to these figures, should more energy have been provided by your isopropanol-based fuel or your ethanol-based fuel?

## 12. Organizing Data
Graph the results of the heating tests for all three fuels on the same graph, with time in minutes on the $x$-axis and temperature in °C on the $y$-axis.

# Conclusions

## 13. Interpreting Graphics
Using your graph from item **12,** which fuel was most effective at heating water? Which was least effective?

## 14. Evaluating Conclusions
Can you think of reasons why using the fuel that burns the fastest and hottest might not be the best choice for this kind of product? (Hint: consider that the fuel is designed to be safely and conveniently carried by campers and hikers.)

## 15. Inferring Conclusions
Ms. LaFaye is concerned about pollution issues related to both the manufacture and the use of the gelled fuel. In particular, she is concerned about the production of pollutants such as nitrogen oxide compounds and hydrocarbons. Are these likely to be produced when the fuel is manufactured or burned? Explain your answers.

# Extensions

## 1. Applying Conclusions
Calculate the amount of temperature change in 75 mL of $H_2O$ per gram of gelled fuel. What mass of each gelled or solid fuel would be necessary to heat a cup (250 mL) of $H_2O$ from 21°C to 100°C?

## 2. Evaluating Methods
If Ms. LaFaye makes fuel chunks just big enough to heat a cup of $H_2O$ to 100°C, as described in Extensions item **1,** what is the unit cost for a chunk of each type of fuel?

| Reagent | Cost |
|---|---|
| Chalk | $6.25/kg |
| Ethanol | $3.27/L |
| Isopropanol | $4.62/L |
| Vinegar | $0.86/L |

# Specific Heat Capacity

## Situation

Today's Housewares, Inc. is planning to introduce a new line of cookware. In deciding which metals to use for the cookware, they have to consider how well the metals retain heat. The metals being considered by the company are aluminum, iron, and copper. Your company has a contract with Today's Housewares, Inc. to test the specific heat capacity of each type of metal. This information, along with other considerations, will be used by Today's Housewares, Inc. to decide which metal to use.

## Background

Heat is a form of energy that cannot be measured directly. However, changes in heat can be determined by measuring changes in temperature. To calculate the amount of heat absorbed or lost by a substance as it changes temperature use the following equation.

$$\text{change in heat} = m \times c_p \times \Delta t$$

Metals with high specific heats tend to heat up and cool down slowly because it takes more energy to raise their temperature.

Measurements in a calorimeter are based on the assumption that the heat absorbed by the water in the calorimeter is the same as the heat released as the metal cools down. However, some heat is lost because it is impossible to insulate perfectly. The problem of heat loss is usually solved by calibrating the calorimeter to determine what adjustments are necessary to compensate for the heat energy that is lost.

## Problem

To get the information for your report to Today's Housewares, Inc., you will need to do the following.
- Make and calibrate a calorimeter using carefully measured masses and temperatures of warm and cold water.
- Measure the mass of each metal sample and heat each one, measuring its temperature.
- Add the heated metal to a carefully measured mass and temperature of cold water in the calorimeter.
- Measure the final temperature of the water and metal to calculate the change in heat energy for the water.
- Determine the specific heat capacity of the metal after accounting for the adjustment specified by the calibration step.

## Objectives

***Build and calibrate*** a simple calorimeter.

***Relate*** measurements of temperature to changes in heat energy.

***Demonstrate*** proficiency in using calorimetry techniques to determine the specific heat capacities of metals.

## Materials

- Aluminum metal sample
- Copper metal sample
- Iron metal sample
- 400 mL beakers, 2
- 100 mL graduated cylinder
- Balance
- Beaker tongs
- Boiling chips
- Glass stirring rod
- Plastic foam cups for calorimeter, 2
- Scissors
- Ring stand
- Test-tube holder
- Test tube, large or medium, 3

**Bunsen burner option**
- Bunsen burner and related equipment
- Ring clamp
- Wire gauze with ceramic center

**Hot plate option**
- Hot plate

**Probe option**
- Thermistor probe

**Thermometer option**
- Thermometer, non-mercury

## Safety

 **Always wear safety goggles and a lab apron to protect your eyes and clothing. If you get a chemical in your eyes, immediately flush the chemical out at the eyewash station while calling to your teacher. Know the locations of the emergency lab shower and eyewash station and the procedure for using them.**

 **When you use a Bunsen burner, confine any long hair and loose clothing. Do not heat glassware that is broken, chipped, or cracked. Use tongs or a hot mitt to handle heated glassware and other equipment because hot glassware does not look hot. If your clothing catches on fire, WALK to the emergency lab shower, and use it to put out the fire.**

 **Scissors are sharp; use with care to avoid cutting yourself or others.**

 **Always clean up the lab and all equipment after use, and dispose of substances according to proper disposal methods. Wash your hands thoroughly before you leave the lab after all lab work is finished.**

## Preparation

1. **Organizing Data**
   Prepare a data table with four columns and four rows for the calibration of your calorimeter. Label the boxes in the top row *Measurement, Cool H₂O, Hot H₂O,* and *Calorimeter.* Label the second through fourth boxes in the first column *Volume (mL), Init. temp. (°C),* and *Final temp. (°C).*

2. **Organizing Data**
   Prepare another data table with four columns and seven rows for the testing of the metals. Label the boxes in the top row *Measurement, Al, Cu,* and *Fe.* Label the second through seventh boxes in the first column *Test-tube mass (g), Metal + test-tube mass (g), Metal init. temp. (°C), H₂O volume (mL), H₂O init. temp (°C),* and *Final temp. (°C).*

3. Construct a calorimeter from two plastic foam cups. Trim the lip of one cup, and use that cup for the top of the calorimeter. Use a pencil to make a small hole in the center of the base of this cup (the top of the calorimeter) so that a thermometer or thermistor can be inserted, as shown in Figure A on the next page. Place the calorimeter in one of the beakers to help prevent it from tipping over.

4. Label three test tubes *Al, Cu,* and *Fe.*

## Technique

### Calibrating the calorimeter

5. Use a graduated cylinder to measure about 50 mL of cold water. Record the exact volume to the nearest 0.1 mL in the calibration data table. (Note: place an "X" in the box on the *Volume (mL)* row in the column labeled *Calorimeter.* The volume of the calorimeter will not be used in this Exploration.)

Nonmercury
thermometer

Thermometer
clamp

Hole in cup
bottom for
thermometer

Trim lip from
this cup
(calorimeter lid)

Leave lip on
this cup
(calorimeter base)

Beaker to keep
calorimeter from
tipping over

**Figure A**

6. Pour the cold water into your calorimeter, and allow the water to reach room temperature. Measure the temperature every minute until the temperature stays the same for 3 min. Record this temperature as the initial temperature for both the cool water and the calorimeter.

7. If you are using a Bunsen burner, clamp a ring to the ring stand, and place the other beaker on the wire gauze on top of the ring clamp. If you are using a hot plate, place the beaker directly on the hot plate. Heat about 75 mL of water in the beaker to 70–80°C.

8. Using beaker tongs, remove the beaker from the heat source, and, using tongs or a hot mitt, pour about 50 mL of hot water into a graduated cylinder. Measure and record the volume to the nearest 0.1 mL in the calibration data table.

9. Using a glass stirring rod, stir the hot water in the graduated cylinder for about a minute and measure the temperature. Record this value in the calibration data table as the initial hot-water temperature. ***Never use a thermometer to stir anything. It will break easily. The glass wall surrounding the bulb is very thin to provide quick and accurate temperature readings.***

10. Immediately pour all of the hot water into the calorimeter with the cool water. Put on the lid and gently swirl the beaker in which the calorimeter rests for 30 s. Record the highest temperature attained by the mixture as the final temperature for the hot water, the cool water, and the calorimeter.

11. Empty the calorimeter. Pour the water down the drain.

***Testing metal samples***

12. Measure the masses of the empty test tubes, and record them in the second data table. Add the appropriate metal to each of the test tubes. The height of the metal in the test tube must be less than the height of the water in the calorimeter cup. Measure and record the mass of each test tube with metal.

**13.** Add about 300 mL of water and a few boiling chips to the beaker from step **7.** Place all three test tubes with metal in the beaker.

Ring stand

Test tube of metal

Beaker of boiling water

Ring clamp with wire gauze

Bunsen burner

Test tube of metal

Beaker of boiling water

Hot plate

**Figure B**

**14.** Heat the beaker of water with the test tubes and metal using a hot plate or a Bunsen burner until the water begins to boil, and continue heating for approximately 10 min as shown in Figure B. Be sure that the metal remains below the surface of the water throughout the heating cycle. Add more water only if necessary.

**15.** While the water boils, measure about 75 mL of cold water in a graduated cylinder. Record the exact volume to the nearest 0.1 mL in the *Al* column in the second data table. Allow the water to come to room temperature, as in step **6.** Record this value as the initial water temperature in the second data table.

**16.** Measure and record the temperature of the boiling water after 10 min. You can assume that each metal is at the same temperature as the boiling water. Record the value in the second data table as the initial temperature for each metal. Keep the water boiling through the following steps.

**17.** First make certain that the thermometer is not in the calorimeter because adding the metal could cause it to break. Then remove the *Al* test tube from the hot-water bath with a test-tube holder because it is hot enough to cause burns. Transfer the aluminum metal to the calorimeter. Quickly and carefully put the top on the calorimeter and gently insert the thermometer into the water. Gently swirl the beaker containing the calorimeter for 30 s, making sure that the metal does not hit the thermometer. Monitor the temperature, and record in your data table the highest temperature attained by the water.

18. Remove the aluminum sample from the calorimeter, and pour the water down the drain. Place the aluminum sample on a paper towel. When it is dry, put it in the appropriate disposal container.

19. Repeat steps 15 and 17–18 for samples of copper and iron.

## Cleanup and Disposal

20. Turn off the boiling water and allow it to cool. Place the dried metals in separate disposal containers designated by your teacher. Clean up your equipment and lab station. After the water has cooled, pour it down the drain. Always wash your hands thoroughly before you leave the lab after cleaning up the lab area and equipment.

## Analysis and Interpretation

1. **Organizing Ideas**
   State the scientific law that is the basis for the assumption that the heat energy lost by the metal as it cools will be equal to the heat energy gained by the water and the calorimeter.

2. **Organizing Data**
   Assuming that 1.00 g/mL is the density of cold water and 0.97 g/mL is the density of hot water, calculate the masses of cold and hot water used in the calibration step.

3. **Analyzing Information**
   Calculate $\Delta t$ for the hot water and the cold water.

4. **Applying Ideas**
   Using the value of the specific heat capacity of water, which is 4.180 J/g·°C, and the specific heat capacity equation given in the Background section, calculate the amounts of heat released by the hot water and absorbed by the cold water.

5. **Evaluating Data**
   Explain why the two values you calculated in item **4** should be slightly different. How much heat energy was released by the calorimeter to the surroundings?

6. **Applying Ideas**
   To be able to use a calorimeter for many different temperature changes without calibrating every time, chemists often determine the heat capacity of the calorimeter, or $C'$, from their calibration data, according to the following equation.

$$C' = \frac{\text{heat lost to surroundings}}{\Delta t_{calorimeter}}$$

   Determine $C'$ for your plastic foam cup calorimeter.

7. **Evaluating Methods**
   Explain whether it is best to have a high or a low value for the heat capacity of a calorimeter.

8. **Organizing Data**
   Calculate the mass of each metal used.

9. **Organizing Data**
   Calculate the temperature changes, $\Delta t$, for the water and for the metal in each of the three tests.

# Conclusions

**10. Organizing Ideas**

Write a valid equation involving these three quantities: heat energy released by the metal as it cools, heat energy absorbed by the water as it warms up, and heat energy released to the surroundings by the calorimeter. (Hint: refer to item **1.**)

**11. Applying Ideas**

Identify which two of the quantities in the equation from item **10** can be calculated from values you already know, have measured, or have calculated in the previous items. (Hint: apply the specific heat capacity equation and the equation in item **6.**)

**12. Applying Ideas**

Set up three different calculations, one for each of the metals tested, to determine the values of the quantities identified in item **11.** Assume that the density of water is 1.00 g/mL.

**13. Inferring Conclusions**

Using the specific heat capacity equation and the answers from the previous items, determine the value of the specific heat capacity for each metal.

**14. Evaluating Conclusions**

Compare your values for specific heats to the values found in a handbook, such as the *CRC Handbook of Chemistry and Physics* or *Lange's Handbook of Chemistry*. (You may need to convert the values given from calories to joules. The conversion factor is: 1 cal = 4.184 J.) Calculate your percent error.

**15. Organizing Ideas**

Refer to your data to determine how the specific heat capacity of the metals compares to the specific heat capacity of water. Which would absorb the same amount of heat energy with a greater change in temperature—metal or water?

**16. Evaluating Conclusions**

Today's Housewares is trying to make pots and pans that will quickly reach cooking temperature. Explain, using your data, which metal is the best choice if this is the primary concern.

# Extensions

**1. Designing Experiments**

Share your data with other lab groups, and calculate a class average for the different specific heat capacities of the metals. Compare the averages to the figures in a chemical handbook, and calculate the percent error for the class averages.

**2. Applying Ideas**

Design a different calorimeter with better insulation. Describe what tests you would use to calibrate it and measure its heat capacity, $C'$. If your teacher approves the design of the calorimeter and your plans for testing it, build the calorimeter and determine how much better it works. Compare this value to the value calculated in the Exploration, and describe the improvement you achieved as a percentage.

**3. Applying Ideas**

Today's Housewares, Inc. is trying to design the handles for its cookware. Should they select materials with high or low specific heat capacities? Explain your answer.

HOLT
ChemFile
LAB PROGRAM

# Specific Heat Capacity—Terrorist Investigation

## TOP SECRET

March 13, 1995

Marissa Bellinghausen
Director of Investigations
CheMystery Labs, Inc.
52 Fulton Street
Springfield, VA 22150

Dear Ms. Bellinghausen:

Yesterday morning, at 0600, CrestAir flight 998 crashed approximately 5 miles from Seymour Johnson AFB, near Goldsboro, NC. Approximately 15 minutes after takeoff from Raleigh-Durham Airport, eyewitnesses heard several loud explosions and saw the plane on fire and hurtling toward the ground. All 50 passengers and 8 crew members were killed on impact.

The Federal Aviation Administration's inspectors have recovered parts of what they believe was an incendiary device placed in the luggage compartment. Reliable sources inform us that several terrorist groups have threatened a wave of weekly airliner bombings. We must identify the responsible party quickly to avert any future calamities.

There are four possible terrorist groups that could have been responsible for this bombing. Each group is known for a specific metal from which they construct their devices.

Popular Front for the Liberation of Fredonia: lead
Penguins' Patrol: zinc
Spleen Activist Group: aluminum
Drachmanian Separatists: iron

Inform us as soon as you can supply evidence to implicate a group.

Sincerely,

*Maxwell Stone*

Maxwell Stone
Director of Anti-Terrorist Operations
North American Theater
Central Intelligence Agency

# Memorandum

**Date:** March 14, 1995
**To:** Thi Antoan
**From:** Marissa Bellinghausen

CheMystery Labs, Inc.
52 Fulton Street
Springfield, VA 22150

For this project, you will be provided with a sample of the metal from the crash site, but it is not very large. Because the CIA hopes to eventually arrest the guilty party, the sample you use must remain intact so that it can be evidence for a later trial. The CIA wants an ironclad case, so check at least one other physical property besides the specific heat capacity that will distinguish these metals. You may need to check reference works for information on the metals that could have been used. Remember, we have less than 72 hours to identify the terrorist group responsible.

The CIA will reimburse us for reasonable expenses, but I want to keep track of how you plan to spend the money, so I need the following items from you before you begin work.

- detailed one-page plan for the procedure, indicating what properties you will measure and how, and all necessary data tables
- detailed list of all equipment and materials you will need, along with individual and total costs

After you complete your work, prepare a two-page report in the form of a letter to Mr. Stone. You must include the following.

- name of the terrorist organization implicated
- summary paragraph describing how you solved this problem
- detailed and organized data and analysis section, including calculations for specific heat of the sample, some other physical property, and averages of multiple trials
- detailed invoice for services and expenses, charged to "Operation Bird-in-the-Hand" (Remember that the invoice will have to get past the CIA's new auditor-general, who is rumored to be a stickler for details.)

## Required Precautions

 Always wear safety goggles and a lab apron to protect your eyes and clothing.

 When you use a Bunsen burner, tie back loose clothing and hair. Do not heat glassware that is broken, chipped, or cracked. Use tongs whenever handling glassware or other equipment that has been heated.

 Do not touch any chemicals. If you get a chemical on your skin or clothing, wash the chemical off at the sink while calling to your teacher.

 Always wash your hands thoroughly when finished.

## Spill Procedures/ Waste Disposal Methods

- Use paper towels to dry the metal samples when you have finished.
- Place the dried metal samples in the containers designated by your teacher.

## References

Review your textbook for information on the specific heat capacities of metals. There are many ways to identify metals based on their physical properties, such as measuring magnetic effects, conductivity, or density. The procedure for measuring specific heat capacity is very similar to a procedure your team recently completed comparing the specific heat capacities of several metals. Your data from that procedure could be useful in helping you identify the metal from the bomb.

### MATERIALS for CIA Operation Bird-in-the-Hand

#### REQUIRED ITEMS
*(You must include all of these in your budget.)*

| | |
|---|---|
| Lab space/fume hood/utilities | 15 000/day |
| Standard disposal fee | 2 000 |
| Balance | 5 000 |
| Beaker tongs | 1 000 |
| Chemical handbook with metal properties | 500 |

#### REAGENTS AND ADDITIONAL EQUIPMENT
*(Include in your budget only what you'll need.)*

| | |
|---|---|
| 100 mL graduated cylinder | 1 000 |
| 400 mL beaker | 2 000 |
| Boiling chips | 500 |
| Bunsen burner | 10 000 |
| Desiccator | 3 000 |
| Filter paper | 500/piece |
| Glass stirring rod | 1 000 |
| Hot plate | 8 000 |
| Magnets | 2 000 |
| Plastic foam cups | 1 000 |
| Ring stand/ring/wire gauze | 2 000 |
| Scissors | 500 |
| Spatula | 500 |
| Stopwatch | 5 000 |
| Test tube, large | 1 000 |
| Test-tube holder | 500 |
| Thermistor probe | 2 000 |
| Thermometer, nonmercury | 2 000 |
| Wash bottle | 500 |
| Weighing paper | 500/piece |

*\* No refunds on returned chemicals or unused equipment.*

#### FINES

| | |
|---|---|
| OSHA safety violation | 2 000/incident |

**Chem**File

# Constructing a Heating/Cooling Curve

## Objectives

**Observe** the temperature and phase changes of a pure substance.

**Measure** the time needed for the melting and freezing of a specified amount of substance.

**Graph** experimental data and determine the melting and freezing points of a pure substance.

**Analyze** the graph for the relationship between melting point and freezing point.

**Identify** the relationship between temperature and phase change for a substance.

**Infer** the relationship between heat energy and phase changes.

**Recognize** the effect of an impurity on the melting point of a substance.

**Analyze** the relationship between energy, entropy, and temperature.

## Situation

Sodium thiosulfate pentahydrate, $Na_2S_2O_3 \cdot 5H_2O$, is produced by a local manufacturing firm and sold nationwide to photography shops, paper processing plants, and textile manufacturers. Purity is one condition of customer satisfaction, so samples of $Na_2S_2O_3 \cdot 5H_2O$ are taken periodically from the production line and tested for purity by an outside testing facility. Your company has been tentatively chosen because your proposal was the only one based on melting and freezing points rather than the more expensive titrations with iodine. To make the contract final, you must convince the manufacturing firm that you can establish accurate standards for comparison.

## Background

As heat energy flows from a liquid, its temperature drops. The entropy, or random ordering of its particles, also decreases until a specific ordering of the particles results in a phase change to a solid. If heat energy is being released or absorbed by a substance remaining at the same temperature, this is evidence that a dramatic change in entropy, such as a phase change, is occurring. Because all of the particles of a pure substance are identical, they all freeze at the same temperature, and the temperature will not change until the phase change is complete. If a substance is impure, the impurities will not lose thermal energy in the same way that the rest of the particles do. Therefore, the freezing point will be somewhat lower, and a range of temperatures instead of a single temperature.

## Problem

To evaluate the samples, you will need a heating/cooling curve for pure $Na_2S_2O_3 \cdot 5H_2O$ that you can use as a standard. To create and use this curve, you must do the following.

• Obtain a measured amount of pure $Na_2S_2O_3 \cdot 5H_2O$.

• Melt and freeze the sample, periodically recording the time and temperature.

• Graph the data to determine the melting and freezing points of pure $Na_2S_2O_3 \cdot 5H_2O$.

• Interpret the changes in energy and entropy involved in these phase changes.

• Verify the observed melting point against the accepted melting point found in reference data from two different sources.

• Use the graph to qualitatively determine whether there are impurities in a sample of $Na_2S_2O_3 \cdot 5H_2O$.

## Safety

Always wear safety goggles and a lab apron to protect your eyes and clothing. If you get a chemical in your eyes, immediately flush the chemical out at the eyewash station while calling to your teacher. Know the locations of the emergency lab shower and eyewash station and the procedure for using them.

Do not touch any chemicals. If you get a chemical on your skin or clothing, wash the chemical off at the sink while calling to your teacher. Make sure you carefully read the labels and follow the directions on all containers of chemicals that you use. Never return leftovers to their original containers; take only small amounts to avoid wasting supplies.

When you use a Bunsen burner, confine any long hair and loose clothing. Do not heat glassware that is broken, chipped, or cracked. Use tongs or a hot mitt to handle glassware or other equipment that has been heated because hot equipment does not always look hot. If your clothing catches on fire, WALK to the emergency lab shower and use it to put out the fire.

Always clean up the lab and all equipment after use, and dispose of substances according to proper disposal methods. Wash your hands thoroughly before you leave the lab after all lab work is finished.

## Materials

- Ice
- $Na_2S_2O_3 \cdot 5H_2O$
- 600 mL beakers, 3
- Balance, centigram
- Beaker tongs
- Chemical reference books
- Forceps
- Graph paper
- Hot mitt
- Plastic washtub
- Ring stands, 2
- Ring clamps, 3
- Ruler
- Stopwatch or clock with a second hand
- Test-tube clamp
- Test tube, Pyrex, medium
- Thermometer clamp
- Wire gauze with ceramic center, 2
- Wire stirrer

**Bunsen burner option**
- Bunsen burner
- Gas tubing
- Striker

**Hot plate option**
- Hot plate

**Probe option**
- Thermistor probes, 2

**Thermometer option**
- Thermometers, non-mercury, 2

## Preparation

**1. Organizing Data**

Prepare spaces in your lab notebook to record the elapsed time and approximate temperature. Prepare a data table in your lab notebook with 6 columns and at least 25 rows. Label the first three boxes in the top row *Time (s), Temp. (°C),* and *Observations of cooling.* Label the fourth through sixth boxes in the top row *Time (s), Temp. (°C),* and *Observations of heating.*

**2.** Fill two 600 mL beakers three-fourths full of tap water.

**3.** Heat water for a hot-water bath. If you are using a Bunsen burner, attach to a ring stand a ring clamp large enough to hold a 600 mL beaker. Adjust the height of the ring until it is 10 cm above the burner. Cover the ring with wire gauze. Set one 600 mL beaker of water on the gauze. If you are using a hot plate, rest the beaker of water directly on the hot plate.

**4.** Monitor the temperature of the water with a thermometer or a thermistor probe. Complete steps **5–8** while the water is heating.

**5.** Cool the water for a cold-water bath. Fill a small plastic washtub with ice. Form a hole in the ice that is large enough for the second 600 mL beaker. Insert the beaker and pack the ice around it up to the level of the water in the beaker.

10 cm piece of wire

Loop that fits into test tube and around thermometer

**Figure A**

6. Bend the piece of wire into the shape of a stirrer, as shown in Figure A. One loop should be narrow enough to fit into the test tube, yet wide enough to easily fit around the thermometer without touching it.

7. Prepare the sample. Assemble the test tube, thermometer, and stirrer, as shown in Figure B. Attach the entire assembly to a second ring stand. Then, add enough $Na_2S_2O_3 \cdot 5H_2O$ crystals so that the test tube is about one-quarter full and the thermometer bulb is well under the surface of the crystals as shown in Figure C.

Stirrer

Thermometer

Test tube

**Figure B**

Thermometer clamp

Thermometer

Stirrer

Test-tube clamp

Test tube

Ring stand

$Na_2S_2O_3 \cdot 5H_2O$ crystals

**Figure C**

8. Set up the container for the hot-water bath as shown in Figure D. Attach two ring clamps, one above the other, to the second ring stand beneath the test-tube assembly. Place a wire gauze with ceramic center on the lower ring. Set a third 600 mL beaker, which should be empty, on the gauze and raise the beaker toward the test-tube assembly until it surrounds nearly one-half of the tube's length. The beaker will pass through the ring clamp without gauze, and the test tube should not touch the bottom or sides of the beaker, as shown in Figure D on the next page. The top clamp keeps the beaker from tipping when the beaker is filled with the hot water.

Figure D

a

b

c

## Technique

### Melting a Solid—Quick Test

9. Check the temperature of the water for the hot-water bath. When it is 85°C, turn off the burner or hot plate. If the temperature is already greater than 85°C, shut off the burner or hot plate, and add a few pieces of ice to bring the temperature down to 85°C. Then, using beaker tongs, remove the beaker of hot water from the burner. Using tongs or a hot mitt carefully pour the water into the empty beaker until the water level is well above the level of the solid inside the test tube. Set the empty beaker on the counter. You will use it again in step **20.**

10. Begin timing. The second the water is poured, one member of the lab group should begin timing, while the other reads the initial temperatures of the bath with one thermometer and sample with the other thermometer.

11. Occasionally stir the melting solid by gently moving the stirrer up and down. Be careful not to break the thermometer bulb. Monitor the temperature of the $Na_2S_2O_3 \cdot 5H_2O$ and the hot-water bath with separate thermometers or probes.

12. When the temperature of the liquid $Na_2S_2O_3 \cdot 5H_2O$ is approximately the same as that of the hot-water bath, stop timing. Record the final temperature of the liquid $Na_2S_2O_3 \cdot 5H_2O$ and the elapsed time in your lab notebook. This temperature is the approximate melting point of your sample. Knowing this value can help you make the careful observations necessary to determine a more precise value.

**Chem**File

13. Using a hot mitt, hold the beaker of hot water with one hand while using the other hand to gently loosen only the lower ring clamp enough so that the beaker of hot water can be lowered and removed. Remove the beaker of hot water, set it on the gauze above the burner, and let it reheat to 65°C while you perform steps **14–20.**

### Freezing a Liquid

14. Set up the cold-water bath. Remove the beaker of cold water from the ice and place it on the ring with the gauze, well below the test tube. Steady the beaker with one hand while raising it until the level of the cold water is well above the level of the liquid inside the test tube. The test tube should not touch the bottom or sides of the beaker.

15. Begin timing. The second that the cold water is in place, one member of the lab group should begin timing, while the other reads the initial temperatures of the sample and the bath. Record the initial time and temperatures in the left half of the data table. The starting temperature of the liquid should be near 80°C.

16. Monitor the cooling process. Measure and record the time and the temperature of the $Na_2S_2O_3 \cdot 5H_2O$ every 15 s in the left half of the data table. Also record observations about the substance's appearance and other properties in the *Observation of cooling* column in the data table. When the temperature reaches 50°C, use forceps to add one or two seed crystals of $Na_2S_2O_3 \cdot 5H_2O$ to the test tube.

17. Continue taking temperature readings every 15 s, stirring continuously, until a constant temperature is attained. (A temperature is constant if it is recorded at four consecutive 15 s intervals.) **Do not try to move the thermometer, thermistor probe, or stirrer when solidification occurs.**

18. Finish timing. Continue taking readings until the temperature of the solid differs from the temperature of the cold-water bath by 5°C.

19. Remove the cold-water bath. Grasp the beaker with one hand, carefully loosen its supporting ring clamp with the other hand, and lower the beaker of cold water away from the test tube. Remove the beaker from the ring and set it on the counter.

### Melting a Solid

20. Set up the container for the hot-water bath. Place the empty beaker from step **9** on the ring and wire gauze. Steady the beaker as you raise it to surround the test tube as you did in step **8,** but this time allow room for the Bunsen burner to be placed under the beaker.

21. Fill the hot water bath. Use the second thermometer or thermistor probe to check the temperature of the water for the hot-water bath. When it is 65°C, turn off the burner or hot plate. If the temperature is greater than 65°C, add a few pieces of ice to lower the temperature. Using tongs or a hot mitt, carefully pour the hot water into the empty beaker until the water level is well above the level of the solid inside the test tube. Set the empty beaker on the counter.

22. Begin timing. The second that the water is poured, one member of the lab group should begin timing while another reads initial temperatures of the water bath and the solid $Na_2S_2O_3 \cdot 5H_2O$. Record the solid's temperature in the right half of the data table. The starting temperature of the solid should be below 35°C.

23. Maintain the bath's temperature. Move the burner or hot plate under the hot-water bath and continue heating the water in the bath. Adjust the position and size of the flame or the setting of the hot plate so that the temperature of the hot-water bath remains between 60°C and 65°C.

24. Monitor the warming process. Record the temperature of the sample every 15 s. Use the stirrer, when it becomes free of the solid, to gently stir the contents of the test tube. Also record observations about the substance's appearance and other properties in the *Observations of heating* column in the data table.

25. Continue taking readings until the temperature of the $Na_2S_2O_3 \cdot 5H_2O$ differs from that of the hot-water bath by 5°C.

26. Record the final temperature and the time.

27. Turn off the burner or hot plate.

## Cleanup and Disposal

28. Remove the thermometer or thermistor probe from the liquid $Na_2S_2O_3 \cdot 5H_2O$ and rinse it. Pour the $Na_2S_2O_3 \cdot 5H_2O$ from the test tube into the disposal container designated by your teacher. If you used a Bunsen burner, check to see that the gas valve is completely turned off. Remember to wash your hands thoroughly after cleaning up the lab area and all equipment.

## Analysis and Interpretation

1. **Organizing Data**
   Plot both the heating and cooling data on the same graph. Place the time (marked in 15 s intervals) on the *x*-axis and the temperature in degrees Celsius on the *y*-axis.

2. **Interpreting Graphics**
   Describe and compare the shape of the cooling curve with the shape of the heating curve.

3. **Interpreting Graphics**
   Locate the freezing and melting temperatures on your graph. Compare them and comment on why they have different names.

4. **Evaluating Methods**
   One purpose of the quick test for melting point is summarized in step **12**. State this purpose and explain how it prepares you for steps **17** and **24**.

5. **Evaluating Data**
   Compare your melting point with that found in references. What is your percent error?

# Conclusions

6. **Analyzing Information**
   As the liquid cools, what is happening to the kinetic energy and the entropy of the following?
   *a.* $Na_2S_2O_3 \cdot 5H_2O$
   *b.* the water bath

7. **Analyzing Information**
   What happened to the temperature of the sample from the time that freezing began until freezing was complete? Did the entropy of the sample increase, decrease, or stay the same?

8. **Predicting Outcomes**
   How would the quantity of the sample affect the time needed for the melting point test?

9. **Predicting Outcomes**
   Would the quantity of the sample used to determine the melting point affect its outcome? (Hint: is melting point an extensive or intensive property?)

10. **Interpreting Graphics**

Examine the graph above, and compare it to your cooling curve. Would this sample of sodium thiosulfate pentahydrate be considered pure or impure? Sketch a line on the graph that represents your cooling data. If the curves are not identical, estimate the difference in melting points.

# Extensions

1. **Interpreting Graphics**
   Refer to the heating and cooling curves you plotted. For each portion of the curve, describe what happens to the energy and entropy of the substance.

## 2. *Applying Ideas*

In northern climates, freezing rain is a driving hazard. When this occurs, warm air from a defroster is blown against the windshield of an automobile in order to restore visibility. It would be convenient to have a system that automatically turned the defroster blower system on and off as needed. A thermostat embedded in the windshield to detect outside temperature could be used to perform this function.

*a.* At what temperature should the thermostat be set to turn on the hot-air blower?

*b.* At what temperature should the thermostat be set to turn off the hot-air blower?

## 3. *Analyzing Methods*

Will crystallization take place if no seed crystal is added? Why or why not?

## 4. *Designing Experiments*

Explain the purpose of a water bath. Why wasn't distilled water necessary?

## 5. *Organizing Ideas*

Which of the following word equations best represents the changes of phase taking place in the situations described in items **1** to **7** below? Place your answer in the space to the left of the numbered items.

*a.* solid + energy $\longrightarrow$ liquid

*b.* liquid $\longrightarrow$ solid + energy

*c.* solid + energy $\longrightarrow$ vapor

*d.* vapor $\longrightarrow$ solid + energy

_____ *1.* ice melting at 0°C

_____ *2.* water freezing at 0°C

_____ *3.* a mixture of ice and water whose relative amounts remain unchanged

_____ *4.* a particle escaping from a solid and becoming a vapor particle

_____ *5.* solids, like camphor and naphthalene, evaporating

_____ *6.* snow melting

_____ *7.* snow forming

_____ *8.* dry ice subliming

_____ *9.* dry ice forming

# Constructing a Cooling Curve—Melting Oils for Soap Production

March 14, 1995

Martha Li-Hsien
Director of Research
CheMystery Labs, Inc.
52 Fulton Street
Springfield, VA 22150

Dear Ms. Li-Hsien:

We are presently in the process of expanding our product line. In the past, we have focused on such items as air fresheners, carpet cleaners, and deodorizing cleansers. We are interested in producing a new line of fresh-smelling body soaps.

As the manufacturing process is developed, we must decide which fat to use in the production of our soap product. We are presented with three possibilities: tallow, vegetable shortening, and coconut oil. There are many factors we must consider, including the energy that the production will require. We would like you to investigate the energy required to melt these fats because soap production requires the fat to be liquid.

We are ready to pay you $175 000 if you give us the following information, which will assist us in deciding on which fat to use.

- a ranking of fats from least to greatest amount of energy required to change phase
- the approximate melting point of each fat
- a cooling curve for each fat
- data from which the cooling curve was graphed

Sincerely,

*Martin Green*

Martin Green
Vice President
Sparkle-Clean Industries

# Memorandum

**Date:** March 15, 1995
**To:** Stacy Logan
**From:** Martha Li-Hsien

CheMystery Labs, Inc.
52 Fulton Street
Springfield, VA 22150

Sparkle-Clean Industries is the kind of client we need. Not only are they leading the market with their products, but they recently sold their dry-cleaning division and have plenty of capital to expand into related product lines. If we do well on this job, we could get more work from them as they test and develop other products as well. My plan is to come in under budget on this project as a way to attract their attention and encourage them to consider us for future work. So be careful about how much money you spend!

Because so much is at stake, I want to be kept informed about your work. Before you begin work, provide the following for me to review.

- a detailed, one-page plan for the procedure and all necessary data tables
- a detailed list of the equipment and materials you will need, along with the individual and total costs

After your preliminary work has been approved, you may begin your lab work. When your research team has completed the lab work, you should prepare a report in the form of a two-page letter to Martin Green at Sparkle-Clean Industries. Mr. Green outlined several requirements for the report in his letter. In addition to those, include the following items.

- a summary paragraph of how you solved this problem
- a discussion of the reliability of the data, including any possible sources of error
- a detailed invoice with costs for equipment, supplies, and services rendered

## Required Precautions

 Always wear safety goggles and a lab apron to protect your eyes and clothing.

 When you use a Bunsen burner, tie back loose clothing and hair. Do not heat glassware that is broken, chipped, or cracked. Use tongs whenever handling glassware or other equipment that has been heated. Do not directly heat the fats, as they are flammable. Instead, use a hot-water bath to heat them.

 Do not touch any chemicals. If you get a chemical on your skin or clothing, wash the chemical off at the sink while calling to your teacher.

 Always wash your hands thoroughly when finished.

## References

The procedure for determining the heating/cooling curve is similar to a procedure you and your team recently completed. Because many fats such as tallow, shortening, and coconut oil are mixtures, not pure substances, it can be difficult to determine the precise melting point. Use at least 10.0 g of each fat so that you will be able to distinguish the melting point. Keep the fats in test tubes because otherwise they will float in the water and will not be evenly heated or cooled. First, heat the fats in a hot-water bath until they are slightly above their melting points (making a heating curve is not necessary). The cooling portion of the experiment can be carried out using a cold-water bath at about 15°C or colder.

## Spill Procedures/ Waste Disposal Methods

- If any of the fats spill, wipe up the spill with a paper towel.
- Dispose of the fats in the separate disposal containers designated by your teacher. You will need to re-heat the fats to transfer them to the disposal containers.
- The equipment used with the fats will need to be scrubbed well, using glassware brushes with plenty of soap and water.

## MATERIALS for Sparkle-Clean Industries
### REQUIRED ITEMS
*(You must include all of these in your budget.)*

| | |
|---|---|
| Lab space/fume hood/utilities | 15 000/day |
| Standard disposal fee | 2 000/g |
| Balance | 5 000 |
| Beaker tongs | 1 000 |

### REAGENTS AND ADDITIONAL EQUIPMENT
*(Include in your budget only what you will need.)*

| | |
|---|---|
| Samples of fats | 500/g |
| 400 mL beaker | 2 000 |
| 100 mL graduated cylinder | 1 000 |
| Bunsen burner/related equipment | 10 000 |
| Buret clamp | 1 000 |
| Filter paper | 500/piece |
| Forceps | 500 |
| Glass stirring rod | 1 000 |
| Hot plate | 8 000 |
| Litmus paper | 1 000/piece |
| Ring stand/ring/wire gauze | 2 000 |
| Scissors | 500 |
| Spatula | 500 |
| Stopwatch | 2 000 |
| Test tube, large | 1 000 |
| Test-tube holder/rack | 2 000 |
| Thermistor probe | 2 000 |
| Thermometer, nonmercury | 2 000 |
| Wash bottle | 500 |
| Watch glass | 1 000 |
| Weighing paper | 500/piece |

\* No refunds on returned chemicals or unused equipment.

### FINES

| | |
|---|---|
| OSHA safety violation | 2 000/incident |

HRW material copyrighted under notice appearing earlier in this work.

HOLT
ChemFile
LAB PROGRAM

# Heat of Fusion

## Situation

The City of Springfield is planning a new sports complex with a convertible arena. The floor of the arena will feature a system that will allow it to be flooded with water and frozen so that it that can be used for hockey or figure skating and then thawed and drained for events such as basketball and gymnastics. The city's recreation department has contacted your company for help in evaluating the cost-effectiveness of the new facility. They need to know how much energy is required to melt the ice so that they can determine how much it will cost each time they change the arena from an ice rink to a gymnasium.

## Background

The freezing and melting points of water and ice are the same temperature: 0°C. The amount of heat energy that must be absorbed to melt 1 mol of solid water at 0°C is the same as the amount of heat energy released as 1 mol of water freezes. This quantity is called the molar heat of fusion. The units for this quantity are kilojoules per mole (kJ/mol). With this value, you can predict the amount of heat energy required to melt any measured quantity of ice.

## Problem

To help the city determine the amount of energy needed to melt an ice rink, you must first determine the molar heat of fusion by doing the following.
- Measure the volume and temperature of some warm water in a calorimeter.
- Add some ice to the warm water, and measure the temperature change and the change in volume after melting is complete.
- Determine how much ice was added.
- Calculate the amount of heat released as the warm water cooled, using the equation for specific heat capacity.
- Calculate the molar heat of fusion of the ice.

## Objectives

*Measure* the volume of a sample of water.

*Calculate* the mass of water given the volume, and then determine the number of moles present.

*Use* a calorimeter to measure the changes in temperature that occur as ice melts.

*Relate* temperature changes to enthalpy changes.

*Calculate* the molar heat of fusion of ice.

## Materials

- Ice
- 400 mL beakers, 2
- 100 mL graduated cylinder
- Balance
- Beaker tongs
- Calorimeter or two plastic foam cups
- Glass stirring rod
- Ring stand

**Bunsen burner option**

- Bunsen burner and related equipment
- Ring clamp
- Wire gauze with ceramic center

**Hot plate option**

- Hot plate

**Probe option**

- Thermistor probe

**Thermometer option**

- Thermometer, non-mercury
- Thermometer clamp

## Safety

 Always wear safety goggles and a lab apron to protect your eyes and clothing. If you get a chemical in your eyes, immediately flush the chemical out at the eyewash station while calling to your teacher. Know the locations of the emergency lab shower and eyewash station and the procedure for using them.

 When you use a Bunsen burner, confine any long hair and loose clothing. Do not heat glassware that is broken, chipped, or cracked. Use tongs or a hot mitt to handle heated glassware and other equipment because hot glassware does not look hot. If your clothing catches on fire, WALK to the emergency lab shower and use it to put out the fire.

 Always clean up the lab and all equipment after use, and dispose of substances according to proper disposal methods. Wash your hands thoroughly before you leave the lab after all lab work is finished.

## Procedure

1. **Organizing Data**

   Prepare a data table in your lab notebook with four columns and five rows. In the first row, label the boxes *Measurement, Trial 1, Trial 2,* and *Trial 3.* In the first column, label the second through fifth boxes *Init. $H_2O$ vol. (mL), Init. $H_2O$ temp. (°C), Final $H_2O$ vol. (mL),* and *Final $H_2O$ Temp. (°C).*

2. If a calorimeter that has already been made is not available, construct one from two plastic foam cups. Trim the lip of one cup and use that cup for the top of the calorimeter. Make a small hole in the base of this cup (the top of the calorimeter) so that a thermometer or thermistor probe can be inserted, as shown in Figure A below. Place the calorimeter in one of the beakers to help prevent it from tipping over.

Nonmercury thermometer

Thermometer clamp

Hole in cup bottom for thermometer

Trim lip from this cup (calorimeter lid)

Leave lip on this cup (calorimeter base)

Beaker to keep calorimeter from tipping over

**Figure A**

# Technique

3. Pour about 350 mL of water into a 400 mL beaker.

4. If you use a Bunsen burner, attach a ring clamp to a ring stand. Place the wire gauze with ceramic center and the beaker on top of the ring clamp. If you use a hot plate, place the beaker directly on the hot plate.

5. Heat the beaker until the water has reached a temperature of about 45°C. Check the temperature every few minutes with a thermometer or a thermistor probe. Do not measure the temperature too close to the surface of the water or too close to the bottom of the beaker.

6. Using beaker tongs or a hot mitt, pour about 100.0 mL of the heated water from the beaker into a graduated cylinder. Measure the volume to the nearest 0.1 mL, and record it in your data table as the initial volume for *Trial 1*.

7. Pour the water into the calorimeter cup, and measure the temperature to the nearest 0.1°C. Record this temperature as the initial temperature for *Trial 1*.

8. Holding the ice cubes with the beaker tongs, shake any excess water off them. Quickly place the ice in the warm water in the calorimeter, and stir the ice water with a glass stirring rod. Place the lid on the calorimeter, and gently insert the thermometer into the hole on the top of the calorimeter lid. Attach the thermometer to the ring stand using the thermometer clamp. Measure the temperature every minute. As soon as the temperature is below 3°C, measure the lowest temperature reached in the next 5 min to the nearest 0.1°C. Record this value in your data table as the final temperature for Trial 1. *Never use a thermometer to stir anything. It will break easily. The glass wall surrounding the bulb is very thin to provide quick and accurate temperature readings.*

9. Remove the calorimeter lid. Using the beaker tongs, remove any unmelted ice. Allow any water on the ice and tongs to drip back into the calorimeter.

10. Using the graduated cylinder, measure the volume of water remaining in the calorimeter to the nearest 0.1 mL. Record this value in your data table as the final volume for *Trial 1*. Because the final total volume will probably be larger than 100.0 mL, it may be necessary to measure the volume of about half of the water in the calorimeter, empty the graduated cylinder, and then measure the volume of the rest of the water.

11. Repeat steps **5–10** twice, recording the results in your data table in the columns for *Trial 2* and *Trial 3*.

# Cleanup and Disposal

12. All water and ice may be washed down the drain. Clean up your equipment and lab station. Always wash your hands thoroughly before you leave the lab after cleaning up the lab area and equipment.

## Analysis and Interpretation

1. **Organizing Data**
   Calculate the volume of water that came from the melted ice for each trial.

2. **Analyzing Data**
   Determine the mass of ice that melted for each trial. (Hint: the density of water is 1.00 g/mL from 0°C to 30°C.)

3. **Analyzing Data**
   Using a periodic table, determine the molar mass of water, and calculate how many moles of water were contained in the ice that melted for each trial.

4. **Analyzing Data**
   Determine the mass of the warm water for each trial. (Hint: the density of water is 0.992 g/mL at 40°C, 0.990 g/mL at 45°C, 0.988 g/mL at 50°C, and 0.986 g/mL at 55°C.)

5. **Organizing Data**
   Calculate the change in temperature, $\Delta t$, of the warm water for each trial.

6. **Organizing Data**
   Calculate the change in temperature, $\Delta t$, of the water that was melted from the ice for each trial. (Hint: assume that the temperature was 0°C initially.)

## Conclusions

7. **Organizing Ideas**
   Write a valid equation relating the following three quantities: the energy absorbed by the ice as it melted, the energy absorbed by the melted water as it warmed, and the energy released by the warm water as it cooled. (Hint: assume that the calorimeter is a perfect insulator. Assume that the ice was at 0°C initially.) Explain the scientific principle that allows you to link these quantities in an equation.

8. **Applying Ideas**
   Identify which two of the quantities in the equation from item **7** can be calculated from information you already have. (Hint: apply the equation for specific heat capacity. Assume that the ice was at 0°C initially.)

9. **Applying Ideas**
   Set up three different sets of calculations, one for each trial, to determine the values of the quantities identified in item **8**. (Hint: the specific heat capacity of water is 4.184 J/g·°C.)

10. **Inferring Conclusions**
    How much heat energy was absorbed by the ice in each of the trials?

11. **Organizing Conclusions**
    On average, how much heat energy is required to melt 1.00 g of ice?

12. **Organizing Conclusions**
    Based on your measurements, what is the molar heat of fusion for water?

### 13. Applying Conclusions

The rink that the city is proposing would be roughly 25 m × 60 m, and the ice would be 2.5 cm thick. If the density of ice is 0.917 g/mL, how much energy would be required to melt this ice?

## Extensions

### 1. Evaluating Conclusions

Compare your values for the molar heat of fusion of water to values that can be found in a handbook such as the *CRC Handbook of Chemistry and Physics* or *Lange's Handbook of Chemistry*. (You may need to convert the values given in the handbook from calories to joules in order to use them. The conversion factor is 1 cal = 4.184 J.) Calculate your percent error.

### 2. Evaluating Conclusions

Calibrate the calorimeter, following the procedure given in the first part of the Technique section in Exploration 5-1. Apply the results to your calculations for this Exploration. Compare these results with those you determined in item **10** of the Conclusions section.

### 3. Designing Experiments

What possible sources of error can you identify with this procedure? If you can think of ways to eliminate them, ask your teacher to approve your plan, and try it out.

### 4. Applying Ideas

A hundred years ago in the United States, vegetables stored for the winter were kept in a "cold cellar." This was an unheated room beneath a house. To prevent these food supplies from freezing during severe winter weather, large barrels of water were placed in the rooms. Explain why this kept the temperature from falling below 0°C.

# Heat of Solution

## Objectives

***Measure*** temperature changes in a calorimeter as a solid solute is dissolved.

***Relate*** temperature changes in a calorimeter to changes in heat energy.

***Determine*** the amount of heat released in joules for each gram of $Na_2S_2O_3$ or $NaC_2H_3O_2$ that dissolves.

***Evaluate*** the use of $Na_2S_2O_3$ and $NaC_2H_3O_2$ in a heating pad.

## Situation

In the summer of 1986, a commercial company began manufacturing an unusual heating pad. It is filled with a liquid solution that exists in a supercooled state at normal temperatures. To use it, a metal disc on the heating pad is pressed and flexed rapidly, causing vibrations that disturb the solution and induce crystallization, which releases a large amount of heat. The heating pad can be reused by placing it in boiling water to remelt the crystals and then slowly cooling it to room temperature. You have been hired by this firm to explore new product possibilities using a solution of sodium thiosulfate, $Na_2S_2O_3$, instead of sodium acetate, $NaC_2H_3O_2$. An important part of this feasibility study is to determine the amount of heat given off per gram of substance, the amount of the substance needed for a heating pad, and the total cost of the substance needed.

## Background

The solution in this heating pad is made by dissolving a salt in water. It is a supersaturated solution; that is, it holds more dissolved salt than is usually possible at a particular temperature. The solution is metastable, so disturbing it causes the ions to become ordered enough to crystallize out of solution, releasing heat energy. The same amount of heat energy is involved when crystals dissolve in water. The bombarding of the crystal lattice by water molecules causes it to break apart. Then these free ions break the hydrogen bonds between water molecules. The water molecules surround the ions, attracted by their charge, and hydrate them. As these interactions take place, energy is released. The sum of the enthalpies of these processes is the heat of solution, $\Delta H_{sol}$. If the hydration step, which involves bond formation, releases more heat than the bond-breaking step, the process is exothermic and $\Delta H_{sol}$ is negative. Otherwise, dissolving is endothermic and $\Delta H_{sol}$ is positive.

## Problem

To compare the cost effectiveness of using $Na_2S_2O_3$ instead of $NaC_2H_3O_2$ in a heating pad, you must do the following.
• Construct a calorimeter and dissolve each salt in it.
• Calculate $\Delta t$ and $\Delta H$ for the water in which the salts dissolve.
• Determine the supplier's price for each salt.
• Calculate $\Delta H$ per gram and the cost-effectiveness for each salt.
• Analyze the precision of your results.

## Safety

Always wear safety goggles and a lab apron to protect your eyes and clothing. If you get a chemical in your eyes, immediately flush the chemical out at the eyewash station while calling to your teacher. Know the locations of the emergency lab shower and eyewash station and the procedure for using them.

Do not touch any chemicals. If you get a chemical on your skin or clothing, wash the chemical off at the sink while calling to your teacher. Carefully read the labels and follow the directions on all containers of chemicals that you use. Do not taste any chemicals or items used in the laboratory. Never return leftovers to their original containers; take only small amounts to avoid wasting supplies.

Always clean up the lab and all equipment after use, and dispose of substances according to proper disposal methods. Wash your hands thoroughly before you leave the lab after all lab work is finished.

## Preparation

**1. Organizing Data**
Prepare a data table with five columns and five rows. Label the boxes in the first row *Measurement, Trial 1—$Na_2S_2O_3$, Trial 2—$Na_2S_2O_3$, Trial 3—$NaC_2H_3O_2$,* and *Trial 4— $NaC_2H_3O_2$.* Label the second through fifth boxes in the first column *Mass of solute (g), Volume of cold $H_2O$ (mL), Initial $H_2O$ temp. (°C),* and *Final $H_2O$ temp. (°C).*

2. Prepare a cold-water bath. Fill a small, plastic washtub with ice. Fill a 600 mL beaker three-fourths full of distilled water. Make a hole in the ice large enough for the beaker. Insert the beaker, and pack the ice around it up to the level of the water.

3. Prepare the calorimeter. Cut a square of corrugated cardboard slightly larger than the top of a 4 oz plastic foam cup. Make a hole in the center of the cardboard piece with a pencil. Make a second hole in the cardboard piece less than 1.0 cm away from the hole in the center. Insert a piece of wire through the hole, and bend each end to make 1.0 cm loops, as shown in Figure A on the next page. Insert a thermometer or a thermistor probe into the center hole and set the entire assembly aside until step **10.**

4. On a piece of weighing paper, measure the mass of approximately 15 g of $Na_2S_2O_3$ to the nearest 0.01 g. Record the mass in your data table.

5. Completely transfer the $Na_2S_2O_3$ to a small Pyrex test tube.

6. Label the tube *Trial 1,* and set it in a test-tube rack.

7. Repeat steps **3–5** for the following.
   **a.** 15 g of $Na_2S_2O_3$; label the test tube *Trial 2*
   **b.** 15 g of $NaC_2H_3O_2$; label the test tube *Trial 3*
   **c.** 15 g of $NaC_2H_3O_2$; label the test tube *Trial 4*

## Materials

- Ice
- $NaC_2H_3O_2$
- $Na_2S_2O_3$
- 600 mL beaker
- 100 mL graduated cylinder
- Balance, centigram
- Corrugated cardboard or lid for cup
- Pencil with point
- Plastic washtub
- Plastic foam cup, small
- Test tubes, small, Pyrex, 4
- Test-tube rack
- Wire, 10–15 cm

**Probe option**
- Thermistor probe

**Thermometer option**
- Thermometer, nonmercury

Wire loop stirrer

Cardboard circle

Thermometer

**Figure A**

## Technique

8. Pour approximately 75 mL of the cold water prepared in step **2** into a 100 mL graduated cylinder. Record the volume to the nearest 0.1 mL.

9. Pour the cold water from the graduated cylinder into the plastic foam cup.

10. Put the thermometer into the cup. The bulb of the thermometer should be completely covered by the water but must not touch the bottom of the cup.

11. Record the temperature of the cold water to the nearest 1.0°C if using a thermometer or to the nearest 0.1°C if using a thermistor probe.

12. Lift the cardboard slightly and dump the contents of the *Trial 1* test tube into the plastic foam cup all at once.

13. Gently move the stirring wire up and down inside the cup to disperse the heat. Allow the solid to dissolve completely.

14. Stir continuously until the temperature of the water peaks.

15. Record the highest temperature reached by the water to the nearest 1.0°C if using a thermometer or to the nearest 0.1°C if using a thermistor probe.

16. Remove, rinse, and dry the thermometer.

17. Pour the solution into the designated waste container.

18. Rinse and dry the cup.

19. Repeat steps **8** through **18** for *Trial 2, Trial 3,* and *Trial 4.*

## Cleanup and Disposal

20. Both substances, $Na_2S_2O_3$ and $NaC_2H_3O_2$, and their solutions, should be placed in their own designated disposal container. Remember to wash your hands thoroughly after cleaning up your lab area and equipment.

## Analysis and Interpretation

1. **Organizing Data**
   Determine the change in temperature, $\Delta t$, of the cold water for each trial.

**Chem**File

EXPLORATION C9 **57**

HRW material copyrighted under notice appearing earlier in this work.

2. **Organizing Information**
   Calculate the heat energy in joules absorbed by the cold water for each trial, using the specific heat capacity equation $\Delta H = \Delta t \times m \times c_p$. Assume the density of water is 1.00 g/mL and the specific heat capacity of water is 4.180 J/g·°C.

3. **Organizing Data**
   Calculate the amount of heat energy released per gram of $Na_2S_2O_3$ for *Trial 1* and *Trial 2*.

4. **Organizing Data**
   Calculate the amount of heat energy released per gram of $NaC_2H_3O_2$ for *Trial 3* and *Trial 4*.

5. **Analyzing Information**
   The accepted value for the heat of solution for $Na_2S_2O_3$ is 200.0 kJ/g. Calculate your percent error.

6. **Analyzing Information**
   Calculate the average of *Trial 1* and *Trial 2*, and the average of *Trial 3* and *Trial 4*. Calculate the percent difference between each trial and the average. Comment on your precision.

7. **Analyzing Information**
   Locate a supplier and find out the cost for $Na_2S_2O_3$ and for $NaC_2H_3O_2$. Calculate the cost per gram for each, and comment on which substance would be the most cost-effective as the primary ingredient of a heat pack.

8. **Evaluating Methods**
   Why was a plastic foam cup instead of a beaker used as a reaction vessel?

9. **Evaluating Methods**
   How could the apparatus for measuring the temperature change of the water be improved?

## Conclusions

10. **Using Information**
    What is the source of the heat energy that is released by the $Na_2S_2O_3$ and absorbed by the water?

11. **Inferring Conclusions**
    What would be the $\Delta H$ of a reaction in which $NaC_2H_3O_2$ crystallizes out of solution?

## Extensions

1. **Inferring Conclusions**
   Why is the fact that the temperature of a phase change remains constant for a pure substance crucial to the development and use of this heating pad?

2. **Relating Ideas**
   When ionic substances dissolve in water, the crystal lattice is broken, and the ions are immediately surrounded by water molecules in a process called hydration. Hydration is exothermic. Breaking the lattice absorbs heat energy. $\Delta H_{sol}$ is the net energy resulting from overcoming the lattice energy and releasing the energy of hydration. Draw a diagram relating $\Delta H_{lat}$, $\Delta H_{hyd}$, and $\Delta H_{sol}$.

# Gas Pressure–Volume Relationship

## Objectives

*Measure* the volume of a gas as pressure is increased.

*Calculate* the pressure on a gas given mass, acceleration due to gravity, and area.

*Graph* the pressure-volume data in two ways.

*Express* the pressure-volume relationship mathematically.

*Calculate* the amount of gas contained in a tank at different pressures.

## Situation

Your company has been contacted to do some research and development work for the Valvopress Company. Valvopress makes valves for a variety of compressed-air tools, and they hope to use this expertise to expand into the scuba-tank market. They need you to help them decide which of several different valves they make would be the best one to install in the new tank. Valvopress has designed a prototype tank that has a volume of 10.5 L, but they need to know, for each of the possible valves, what volume of air at 1.0 atm of pressure could be compressed into a fully pressurized tank.

## Background

Boyle's law describes the relationship between the pressure and the volume of a gas at constant temperature. You will make measurements to test Boyle's law using a sample of air enclosed in the tube of a Boyle's law apparatus. After measuring the volume at normal pressure, you will measure how the volume changes as pressure increases.

Pressure is equal to the amount of force exerted on an area. You will increase the pressure by stacking objects on top of a plunger. Because of gravity, the objects will exert a force (in N) equal to 9.801 $m/s^2$ times the value of their mass (in kg).

Remember that normal pressure at sea level is defined to be exactly 1 atm. Unless your teacher instructs you otherwise, use this value as the normal pressure in your work. Remember:
1.00 atm = 29.92 in of Hg = 760 mm of Hg = $1.013 \times 10^5$ Pa = 101.3 kPa.

## Problem

To evaluate the different valves, your company needs to do the following.
- Vary the pressure on a fixed amount of air and measure the volume.
- Plot graphs of pressure versus volume and pressure versus the inverse of volume.
- From the graphs, determine how pressure and volume are related.
- From the relationship between pressure and volume, calculate what volume of air at normal pressure is equivalent to a tankful of pressurized air for each of the valves.

## Safety

Always wear safety goggles and a lab apron to protect your eyes and clothing. If you get a chemical in your eyes, immediately flush the chemical out at the eyewash station while calling to your teacher. Know the locations of the emergency lab shower and eyewash station and the procedure for using them.

Always clean up the lab and all equipment after use, and dispose of substances according to proper disposal methods. Wash your hands thoroughly before you leave the lab after all lab work is finished.

## Preparation

1. **Organizing Data**
   Prepare a data table with six columns and five rows. In the first through fifth rows of the first column, write the following labels: *Masses added, Mass (g), Trial 1 vol. (cm³), Trial 2 vol. (cm³)*, and *Trial 3 vol. (cm³)*. In the second through sixth columns of the first row, write these labels: *0 mass, Mass 1, Mass 2, Mass 3*, and *Mass 4*. You will also need a space to record the diameter of the syringe barrel below the data table.

2. The syringe or your Boyle's law apparatus is marked in units of cubic centimeters, abbreviated cc or cm³. Volume in cubic centimeters is the same as volume in milliliters.

3. Using removable Post-It notes, label the objects that will be stacked on the plunger of the syringe *1* through *4*.

## Technique

4. Using a balance, measure the mass of each of the four objects you will use as masses, with their labels attached. Record each mass in the appropriate space in the data table. Under the column labeled *0 mass*, fill in a zero.

5. Using a metric ruler, record the diameter of the syringe or Boyle's law apparatus to the nearest millimeter, and record it in your lab notebook.

6. Set up the syringe as your teacher instructs so that you will be able to place the masses on the plunger without the apparatus falling over. If you are using a Boyle's law apparatus, this step is not necessary.

7. If you are using a Boyle's law apparatus, trap about 35 cm³ of air in the syringe of the Boyle's law apparatus. If you are using a sealed syringe, trap about 9 cm³ of air in the syringe. For either piece of equipment, pull the plunger all of the way out of the syringe. Insert a piece of carpet thread into the barrel with the end of the thread hanging out to make a small space for air to escape, as shown in Figure A on the following page. Push the plunger back into the syringe and position the head at the appropriate location, either 35 cm³ for a Boyle's law apparatus or 9 cm³ for a sealed syringe.

## Materials

- Balance
- Carpet thread
- Removable Post-It notes
- Ruler, metric
- Sealed syringe or Boyle's law apparatus
- Stackable objects of roughly equal mass, such as books or blocks of wood (500–550 g each), 4

Carpet thread (used only to adjust volume of trapped air)

**Figure A**

8. While holding the plunger in place, carefully pull out the thread. Twist the plunger several times to allow the head to overcome any frictional forces. Read the volume to the nearest 0.1 cm³, and record this value in your data table as the initial volume for zero mass.

9. Place the first of the masses on the plunger. Your partner should stand by to ensure that the masses do not fall. Give the plunger several twists to reduce frictional forces. When the plunger comes to rest, read and record the volume to the nearest 0.1 cm³ in the row for *Trial 1*.

10. Carefully position the second mass on top of the first. Your partner should stand by to ensure that the masses do not slip off as you twist the plunger. Allow the plunger to come to rest before you read and record the volume.

11. Follow the procedure described in step **10** as you add the third and fourth masses. Record the volume after each addition.

12. Remove all four masses and repeat the process two more times, starting with step **9**, recording the results in the rows for *Trial 2* and *Trial 3*.

## Cleanup and Disposal

13. Clean all apparatus and your lab station. Remember to wash your hands.

## Analysis and Interpretation

1. **Organizing Data**
   By adding the masses you measured, calculate the total mass that was on the plunger for each of the volumes measured (zero to four masses).

2. **Analyzing Data**
   Determine the cross-sectional area of the plunger using the diameter recorded in the data table. Then convert your answer into units of square meters, m².

   $$\left(\text{Hint: for a circle, area} = \pi\left(\frac{\text{diameter}}{2}\right)^2\right)$$

## 3. Analyzing Data

The force in newtons that masses exert due to gravity is equal to their mass in kilograms times the value of their acceleration due to gravity. For objects near the Earth's surface, the acceleration due to gravity, $g$, has a value of 9.801 m/s². Calculate the force exerted by the masses on the plunger for each of the volumes measured.

$$F = mg = m \times 9.801 \text{ m/s}^2$$

## 4. Organizing Data

Pressure is the force exerted divided by the area over which the force is exerted. If the force is in units of newtons and the area is in units of square meters, the pressure calculated will be in units of pascals (Pa). Calculate the pressure exerted by the masses for each of the volumes measured.

## 5. Resolving Discrepancies

Why would it be incorrect to say that there is no pressure on the trapped air when no masses are on the plunger? What is the total pressure exerted on the sample before any masses have been added to the plunger? (Hint: it is not equal to zero.)

## 6. Analyzing Data

Using your answers to items **4** and **5**, calculate the total pressure for each stage of the procedure.

## 7. Organizing Data

Calculate the average volume you measured in the three trials for each of the five volumes measured.

## 8. Analyzing Data

Examine your data table. What conclusions can you draw about the relationship of pressure to volume?

## 9. Interpreting Graphics

Make a graph of pressure versus volume. Plot pressure on the horizontal axis and average volume on the vertical axis. Draw a smooth curve through the points.

## 10. Organizing Data

Calculate 1/V, the inverse of each of the average volumes. For example, if the average volume is 26.5 cm³, then $1/V =$

$$\frac{1}{26.5 \text{ cm}^3} = 0.0377 \text{ cm}^{-3}.$$

## 11. Interpreting Graphics

Make a graph of pressure versus 1/V. Plot pressure on the horizontal axis and 1/V on the vertical axis. Draw the best-fit line through the points.

## 12. Interpreting Graphics

A straight line has an equation of the form $y = mx + b$. From your graph, what are the values of $m$ and $b$ if 1/V and $P$ are substituted for $y$ and $x$, respectively? (Hint: if you have a graphics calculator, use the $\boxed{\text{STAT}}$ mode to enter your data and make a linear regression equation using the LinReg function from the STAT menu.)

## 13. Analyzing Methods

Explain how you can be certain that neither the temperature nor the number of moles of gas changed during the course of the procedure.

## 14. Analyzing Methods

What would have happened if there had been no carpet thread in step **7** and you tried to adjust the plunger?

# Conclusions

### 15. Inferring Conclusions

Rearrange your equation from item **12** so that $P$ and $V$ are on the same side of the equation. Substitute the symbol $k$ for the constant term. (Hint: the approximation of very small amounts as being equal to zero may be necessary.)

### 16. Evaluating Conclusions

Consider the value of $k$ that was found for the sample of gas in the procedure. Can you use the same value of $k$ for pressure and volume calculations for the scuba tank? (Hint: consider what variables were held constant during the procedure.)

### 17. Applying Conclusions

Using the equation from item **15,** how can you relate these four quantities: the pressure in the scuba tank, the volume of the air in the tank at that pressure, normal atmospheric pressure, and the volume of the air in the scuba tank if it was at normal atmospheric pressure.

### 18. Applying Ideas

Using your answer to item **17,** calculate the volume of air at normal atmospheric pressure that could be contained in a scuba tank at the maximum pressure for each of the valves listed in the table shown below.

| Valve Serial Number | Top Pressure |
|:---:|:---:|
| PDE-57 | 112 atm |
| RRM-82 | 168 atm |
| RMD-332 | 219 atm |
| ZRH-61 | 240 atm |
| XJP-27 | 253 atm |

### 19. Applying Ideas

For every 1000 L of air at normal atmospheric pressure, a diver can spend, on average, 26 min underwater at a depth of 20 ft. Using your answer to item **18,** calculate how long you could stay underwater with tanks made using each of the valves.

### 20. Evaluating Viewpoints

The unit costs for the different possible components of the scuba tank are given in the following table. What will be the unit cost of the most efficient scuba tank, in terms of volume of gas at normal atmospheric pressure compared to cost?

| Item | Cost |
|---|---|
| Filling tank with air | $ 1.50 |
| External pressure meter | $ 20.00 |
| Tank | $ 50.00 |
| PDE-57 valve | $ 37.50 |
| RRM-82 valve | $ 40.00 |
| RMD-332 valve | $ 45.00 |
| ZRH-61 valve | $ 70.00 |
| XJP-27 valve | $105.00 |

## Extensions

**1. Connecting Ideas**

Explain how the relationship you observed between the pressure and volume of a gas is consistent with the kinetic molecular theory of gases.

**2. Evaluating Methods**

The value of the pressure that you applied to the gas in the form of masses on the plunger of the syringe is referred to as the value of the pressure of the gas itself. Explain why this is logical.

**3. Designing Experiments**

What possible sources of error can you identify in your procedure? If you can think of ways to eliminate them, ask your teacher to approve your suggestions, and run the procedure again.

**4. Applying Conclusions**

Imagine a gas confined in an unbreakable container. In theory, the pressure on the gas could be increased gradually until it approached infinity. What happens to the volume of the gas as the pressure becomes infinitely high?

**5. Research and Communication**

Find out about the original experiment that Robert Boyle performed to establish the relationship you just explored. Write a short paper describing his experimental setup and procedure. Include a sketch of the equipment he used, and compare it to the Boyle's law apparatus you used.

# Gas Temperature–Volume Relationship: Balloon Flight

March 15, 1995

Ms. Sandra Fernandez
Director of Development
CheMystery Labs, Inc.
52 Fulton Street
Springfield, VA 22150

Dear Ms. Fernandez:

Our corporation, Barotherm, Inc. is developing a new type of hot-air balloon. Instead of using gas heaters, we plan to use electric heaters powered by new flexible solar panels on the balloon's top. In this way, the balloon can be kept sealed.

We are also creating an "autopilot" system to control the balloon's flight using temperature sensors sewn into the fabric of the balloon. In order to program the system's computer, we need you to determine the internal temperature necessary to keep the balloon aloft.

The balloon can hold as much as $2.6 \times 10^3$ m³ of air, but it may not be necessary to fully inflate the balloon to achieve the proper amount of lift. Before heating, the balloon will be sealed at 15°C with $2.0 \times 10^3$ m³ of air inside. The mass of the balloon's equipment and payload will be 480 kg.

To prepare for these calculations, experimentally test the relationship between temperature and volume for a sample of air. Any deviations from ideal gas behavior should be incorporated into your calculations.

We are willing to pay $100 000 for this work, provided it can be completed in a timely fashion so that we can begin testing the prototype model as soon as possible.

Sincerely

*Vitra Muhmedi*

Vitra Muhmedi

# Memorandum

*CheMystery Labs, Inc.*
*52 Fulton Street*
*Springfield, VA 22150*

**Date:** March 16, 1995
**To:** Leah Quarters
**From:** Sandra Fernandez

As you know, according to Charles's law, an ideal gas sample will have a volume proportional to the temperature in kelvins. But real gases such as air can show some deviations from this ideal behavior. Thus, it is necessary for us to test samples of air ourselves.

I propose using a setup like the one sketched here. Before you begin the work, I need the following.

- detailed one-page plan and all necessary data tables
- detailed list of all equipment and materials you will need, along with individual and total costs
- well-organized calculations for the necessary balloon temperature, assuming air is an ideal gas

Thermometer or LEAP System thermistor probe

Sealed syringe

Plunger, set at 5.0 cm³ and lubricated with a little mineral oil or glycerin

Hot-water bath, at about same level as plunger

Bunsen burner or hot plate

The balloon will begin to float when the combined mass of the payload, the balloon, and the heated air inside it is less than the mass of the air displaced by the balloon's volume. Some necessary information is given in the References section.

When you have completed your work, prepare a report in the form of a two-page letter to Ms. Muhmedi at Barotherm. Be sure to cover the following points.

- temperature and volume necessary for the balloon to float, along with calculations
- brief description of the procedure
- detailed and organized data section
- organized analysis section explaining and graphing the data, as well as any possible sources of error
- detailed invoice for services and expenses

## Required Precautions

 Always wear safety goggles and a lab apron to protect your eyes and clothing.

 Tie back loose clothing and hair. Do not heat glassware that is broken, chipped, or cracked. Use tongs whenever handling glassware or other equipment that has been heated because hot glassware does not always look hot.

 If you will be using a thermometer, be sure that you are using a thermometer clamp to hold it in place or that the thermometer has already been inserted into a split one-hole stopper. Clamp the stopper, not the thermometer itself.

 Always wash your hands thoroughly when finished.

## Spill Procedures/ Waste Disposal Methods

- Water from the hot-water bath may be poured down the drain.
- Be sure to thoroughly clean the plunger of the syringe with soap and water.

## References

Review Charles's law, which relates temperature and volume for gases.

On the ground, the density of air at 15°C and normal pressure is 1.22 g/L. At the balloon's flying altitude of 500 m, the temperature drops slightly, and the density of air is 1.17 g/L.

## MATERIALS for Barotherm, Inc. balloon project

### REQUIRED ITEMS
(You must include all of these in your budget.)

| | |
|---|---|
| Lab space/fume hood/utilities | 15 000/day |
| Standard disposal fee | 2 000/g of product |
| Beaker tongs | 2 000 |

### REAGENTS AND ADDITIONAL EQUIPMENT
(Include in your budget only what you'll need.)

| | |
|---|---|
| Lubricant (mineral oil or glycerin) | 1 000/mL |
| 400 mL breaker | 2 000 |
| 100 mL graduated cylinder | 1 000 |
| 10 cm³ sealed syringe | 5 000 |
| Bunsen burner/related equipment | 10 000 |
| Clamps | 1 000 |
| Filter paper | 500/piece |
| Glass funnel | 1 000 |
| Glass stirring rod | 1 000 |
| Hot plate | 8 000 |
| Ring stand/ring/wire gauze | 2 000 |
| Six test tubes/holder/rack | 2 000 |
| Thermistor probe | 2 000 |
| Thermometer, nonmercury, in one-hole stopper | 2 000 |
| Wash bottle | 500 |
| Weighing paper | 500/piece |

* No refunds on returned chemicals or unused equipment.

### FINES

| | |
|---|---|
| OSHA safety violation | 2 000/incident |

# Masses of Equal Volumes of Gases

## Situation

Compressed Gases, Inc. provides gases to customers in heavy steel tanks. For safety's sake, tanks are filled and refilled with only one gas, and a different color is used for tanks of each gas. At the factory of one customer, an overenthusiastic painter painted several different tanks the same color. The customer believes that the tanks contained carbon dioxide, $CO_2$, and methane, $CH_4$, but isn't sure. The customer also uses hydrogen, $H_2$, and oxygen, $O_2$. In their pure form, all of these gases are odorless and colorless, so they are hard to distinguish.

## Background

Avogadro's principle states that equal volumes of gas, measured at the same temperature and pressure, contain the same number of molecules. Thus, if you measure the masses of these equal volumes, you can be sure that you have measured the masses of equal numbers of molecules. In other words, the ratio is the same as the mass ratio of a molecule of $CO_2$ to a molecule of $CH_4$.

However, when the mass of an object is measured in a fluid such as air, the fluid buoys up the object, making its mass appear to be less. The difference between the apparent and actual mass is equal to the mass of the air that is displaced by the object's volume. For most solid and liquid objects, the difference is small. But for a gas, its mass is close to the mass of the air it displaces. To find the actual mass, the mass of the air displaced by the gas must be added to the apparent mass of the gas. Suppose the volume of a gas is exactly 1.00 L at 20°C and 760 mm Hg. At this temperature and pressure, the density of air is 1.2 g/L. The apparent mass of the gas will be 1.2 g less than if it was measured in a vacuum. To obtain the actual mass of the gas, 1.2 g must be added to its apparent mass.

## Problem

The staff at Compressed Gases, Inc. determined the ratio of the masses of the two gases as 2.0 : 1.0. In order to verify the identities of the gases, you must do the following.

- Experimentally verify this mass ratio at constant temperature and pressure using known samples of pure $CO_2$ and pure $CH_4$.

- Compare your results to those of Compressed Gases, Inc. to decide whether the tanks are likely to contain $CO_2$ and $CH_4$.

- If so, explain which tank contains which gas.

- If not, explain which gases the tanks do contain.

## Objectives

***Measure*** the masses of equal volumes of methane and carbon dioxide under the same conditions of temperature and pressure.

***Measure*** the volume of the gas in the plastic bag by water displacement.

***Determine*** the actual masses of the gases by eliminating the effect of buoyancy.

***Calculate*** the ratio of the two masses.

***Apply*** Avogadro's principle to determine the identity of gases by determining the ratios of their molar masses.

## Materials

- Dry ice
- Methane gas (from gas jet)
- 2 L glass bottle
- 250 mL or 500 mL graduated cylinder
- 1 L plastic bag
- Balance
- Crucible tongs
- Medicine dropper inserted in a rubber stopper
- Pinch clamp
- Plastic straw
- Pneumatic trough
- Rubber band
- Rubber tubing, 50 cm

## Safety

 **Always wear safety goggles and a lab apron to protect your eyes and clothing. If you get a chemical in your eyes, immediately flush the chemical out at the eyewash station while calling to your teacher. Know the locations of the emergency lab shower and eyewash station and the procedure for using them.**

 **Do not touch any chemicals. If you get a chemical on your skin or clothing, wash the chemical off at the sink while calling to your teacher. Make sure you carefully read the labels and follow the directions on all containers of chemicals that you use. Do not taste any chemicals or items used in the laboratory. Never return leftovers to their original containers; take only small amounts to avoid wasting supplies.**

 **Methane is flammable, and mixtures of methane and air can explode if ignited. Make sure there are no flames anywhere in the room while anyone is working with methane. Do not turn any electric switches on or off while anyone is working with methane. When you finish filling the bag, check twice to be sure that your gas valve and the next closest one are turned off.**

 **Never handle dry ice with your hands; it is cold enough to cause burns. Use tongs to pick it up, or wear well-insulated cloth gloves when you touch it. Wrap the medicine dropper in a layer of cloth and wear disposable plastic gloves when fitting the medicine dropper into the tubing.**

 **Carbon dioxide is an asphyxiant. If it is stored in a container such as an ice chest, do not put your head near it.**

 **Always clean up the lab and all equipment after use, and dispose of substances according to proper disposal methods. Wash your hands thoroughly before you leave the lab after all lab work is finished.**

## Preparation

**1. Organizing Data**

Prepare a data table with spaces for the following: the room temperature and pressure in the lab, the volume of the bag, the mass of the bag assembly, and the mass of the bag assembly when filled with $CO_2$ and with $CH_4$.

## Technique

**2.** The plastic bag you use should be dry and free of holes. Fold or roll the bag so that there is no air in it. Measure the total mass of the bag, a rubber band, a rubber stopper, and a medicine dropper. Record this mass (the mass of the bag assembly) to the nearest 0.01 g in your data table.

**Chem**File

EXPLORATION C11 · **69**

HRW material copyrighted under notice appearing earlier in this work.

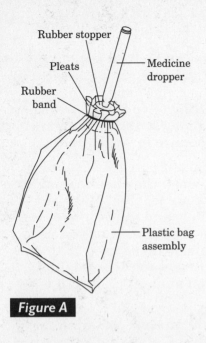

Rubber stopper

Pleats

Rubber band

Medicine dropper

Plastic bag assembly

**Figure A**

3. Using tongs, obtain a piece of dry ice about the size of a large pea. Place the dry ice in the bag.

4. As the dry ice sublimes to form gaseous $CO_2$, gather the open end of the plastic bag in small pleats around the wide end of the rubber stopper. Secure the bag tightly with a rubber band.

5. Handle the bag assembly by the rubber stopper only. Keep the bulb of the medicine dropper off and the dropper pointed up, as shown in Figure A, so that any air in the bag will be expelled as the more dense $CO_2$ fills the bag from the bottom.

6. When all of the dry ice has sublimed and the pressure in the bag is equal to the pressure of the room, the bag will stop expanding. Replace the rubber bulb on the medicine dropper. Measure the mass of the bag and $CO_2$ to the nearest 0.01 g. Record the mass in your data table.

7. Roll up the bag from the bottom to force out all of the carbon dioxide gas. Following the precautions for hand safety, use a length of rubber tubing to connect the gas jet to the medicine dropper. Open the gas valve slightly to fill the bag with $CH_4$ from the gas jet. Then turn the gas valve off. In the fume hood, allow the pressure inside the bag to stabilize, and then replace the bulb on the medicine dropper. The bag should be inflated to the same volume as when the $CO_2$ mass was measured. Obtain the mass of the bag assembly and $CH_4$ to the nearest 0.01 g and record the mass in your data table.

8. In the fume hood or outdoors, roll up the bag to expel the $CH_4$.

9. Use a plastic straw to blow air into the bag through the medicine dropper. Allow the pressure of the air to become equal to atmospheric pressure. The bag's volume should be the same as when it was inflated with $CO_2$ and $CH_4$. Following the precautions for hand safety, attach a piece of rubber tubing to the dropper and fasten a pinch clamp on the rubber tubing approximately 5 cm from the end of the dropper.

10. Completely fill a 2 L bottle with tap water. Also fill the pneumatic trough with tapwater to the overflow spout. Position the overflow spout over a drain. Place your hand over the mouth of the bottle, invert it, and lower the mouth of the bottle below the surface of the water. Remove your hand from the bottle, and rest the inverted bottle on the shelf. While your lab partner supports the bottle, insert the end of the rubber tubing into the bottle, as shown in Figure B on the next page.

Plastic bag assembly

2 L glass bottle

Pneumatic trough

**Figure B**

11. Hold the bag tightly around the rubber stopper to ensure a good seal. Lift the bag above the level of the water and remove the pinch clamp. Very gently squeeze the gas out of the bag. As the bag deflates, crumple or roll it until all of the gas has been removed. Immediately close off the rubber tubing with a pinch clamp and pull the tubing out of the bottle.

12. Place your hand under the water, over the opening of the bottle. Invert the bottle and place it on the lab desk. Use a large graduated cylinder to determine the volume of water required to refill the bottle. Record this volume to the nearest milliliter as the volume of the bag.

13. Record the temperature and pressure in the room.

## Cleanup and Disposal

14. Clean all apparatus and your lab station. Make sure that the gas valve at your lab station and the one next to your station are completely shut off before leaving the laboratory. Remember to wash your hands.

## Analysis and Interpretation

1. **Evaluating Data**
   Were the masses measured in this lab actual or apparent masses? Explain your answer.

2. **Organizing Data**
   What portion of the mass measured for $CO_2$ and the assembly was due to $CO_2$? What portion of the mass measured for $CH_4$ and the assembly was due to $CH_4$?

3. **Organizing Data**
   Use the table on the following page to find the density of air at the temperature and pressure of your lab. Calculate the mass of air displaced by the filled bag assembly. The mass of air displaced is equal to the density of air multiplied by the volume of the bag.

| Density of Air g/L | | | | |
|---|---|---|---|---|
| Pressure (mm Hg) | 15°C | 20°C | 25°C | 30°C |
| 720 | 1.16 | 1.14 | 1.12 | 1.10 |
| 730 | 1.18 | 1.16 | 1.14 | 1.12 |
| 740 | 1.19 | 1.17 | 1.15 | 1.13 |
| 750 | 1.21 | 1.19 | 1.17 | 1.16 |
| 760 | 1.23 | 1.20 | 1.18 | 1.16 |
| 770 | 1.24 | 1.22 | 1.20 | 1.18 |

4. **Organizing Data**
Using your answers to items **1** and **3** and the information given, calculate the actual mass of the $CO_2$ and $CH_4$ samples.

5. **Organizing Data**
Determine the ratio of the actual mass of $CO_2$ to the actual mass of $CH_4$.

## Conclusions

6. **Evaluating Conclusions**
Calculate the molar mass for $CO_2$, $CH_4$, and the theoretical mass ratio for $CO_2$ and $CH_4$.

7. **Evaluating Methods**
Calculate the percent error for your experimental determination of the mass ratio for $CO_2$ and $CH_4$.

8. **Evaluating Conclusions**
Based on your results, is it likely that the tanks, whose contents had a mass ratio of 2.0 : 1.0, contain $CO_2$ and $CH_4$? (Hint: consider whether or not your experimental range of error includes the values 2.0 : 1.0 or 1.0 : 2.0.)

9. **Applying Ideas**
Using the periodic table, calculate all possible mass ratios for the following gases: $H_2$, $CH_4$, $O_2$, $CO_2$. What are the likeliest possible gases contained in the tanks?

## Extensions

1. **Evaluating Methods**
Why was it important that you obtain the mass of the methane and carbon dioxide at the same temperature and pressure?

2. **Designing Experiments**
What possible sources of error can you identify in this procedure? If you can think of ways to eliminate them, ask your teacher to approve your suggestions, and run the procedure again.

3. **Applying Ideas**
A student found that the actual mass of a sample of an unknown gas was 1.30 g. A sample of oxygen gas of the same volume at the same temperature and pressure had a mass of 1.48 g. What is the molar mass of the unknown gas?

HOLT
ChemFile
LAB PROGRAM

# Paper Chromatography

## Objectives

***Demonstrate*** proficiency in qualitatively separating mixtures using paper chromatography.

***Compare*** inks by using paper chromatography with a variety of solvents.

***Evaluate*** samples to establish which pen was used on a document.

## Situation

Recently, handwriting experts discovered a set of forgeries. Several museums in the United States had been displaying documents supposedly signed by Abraham Lincoln. The FBI suspects that this could be the work of Benny "Fingers" Smithson, who was recently paroled from prison after serving time for his part in a phony Babe Ruth autograph scam. Smithson denies his involvement. A search warrant was issued, and the FBI found three pens in his apartment. The FBI is also investigating another suspect, Thomas Banks, an employee of one of the museums. Three pens from his belongings are also being held as evidence. Before the FBI can press charges, they need conclusive evidence linking the pens and the phony signatures.

## Background

Paper chromatography is a method of separating mixtures by using a piece of absorbent paper. In this process, the solution to be separated is placed on a piece of dry filter paper (the stationary phase). A solvent (the moving phase) is allowed to travel across the paper by capillary action. As the solvent is soaked up by the paper, some of the components of the mixture are carried with it. The components of the mixture that are most soluble in the solvent and least attracted to the paper travel the farthest. The resulting pattern of molecules is called a *chromatogram*. In cases where the molecules are easily visible such as in inks, this method distinguishes the components of a mixture.

## Problem

To determine which pen was used in the forgery, you must do the following.

- Prepare chromatograms for each of the pens using the two different solvents.
- Prepare chromatograms for different parts of the forged signature using two different solvents.
- Compare the chromatograms and decide which pen is the likeliest match.
- Provide specific examples of similarities between chromatograms, citing measurable points of comparison.

## Safety

Always wear safety goggles and a lab apron to protect your eyes and clothing. If you get a chemical in your eyes, immediately flush the chemical out at the eyewash station while calling to your teacher. Know the locations of the emergency lab shower and eyewash station and the procedure for using them.

Do not touch any chemicals. If you get a chemical on your skin or clothing, wash the chemical off at the sink while calling to your teacher. Make sure you carefully read the labels and follow the directions on all containers of chemicals that you use. Never return leftovers to their original containers; take only small amounts to avoid wasting supplies.

Because the isopropanol is flammable, no Bunsen burners, hot plates, or other sources of heat should be in use in the room during this lab. Carry out all work with isopropanol in a hood.

Always clean up the lab and all equipment after use, and dispose of substances according to proper disposal methods. Wash your hands thoroughly before you leave the lab after all lab work is finished.

## Materials

- Distilled water
- Isopropanol
- Filter papers, 12 cm, 4
- Filter paper wicks, 2 cm equilateral triangles, 2
- Forged signature samples
- Paper clips, 2
- Pens, black ink, 6
- Pencil
- Petri dish with lid
- Ruler

## Preparation

**I. Organizing Data**

Prepare your data table with seven columns and eight rows. Label the boxes in the top row *Pen number, H₂O—center, H₂O—middle, H₂O—edge, Isopropanol—center, Isopropanol—middle,* and *Isopropanol—edge.* In the second through eighth columns of the first column, add the labels *Smithson 1, Smithson 2, Smithson 3, Banks 4, Banks 5, Banks 6,* and *Forgery.*

## Technique

2. Use a pencil to sketch a circle about the size of a quarter in the center of the piece of filter paper. Write the numbers 1–6 in pencil around the inside of this circle, as shown in Figure A on the next page.

3. On the circle beside the number 1, use pen number 1 to make a large dot. Use pen number 2 to make a dot beside number 2, and repeat this procedure for each pen.

4. Repeat steps **2** and **3** with a second piece of filter paper. One will be used with water as a solvent, and the second will be used with isopropanol as a solvent.

5. Roll up the triangle of filter paper to be used as a wick. Use the pencil to poke a small hole in the center of the first marked piece of filter paper. Insert a rolled-up piece of the wick through the hole, as shown in Figure B on the next page.

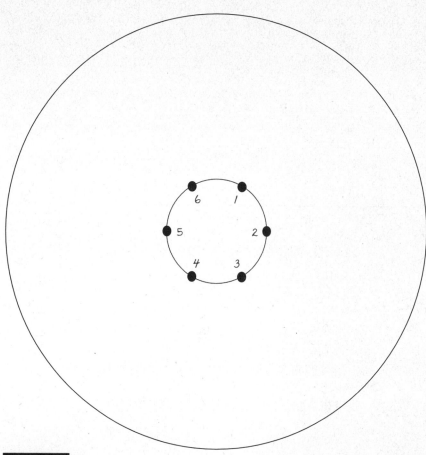

**Figure A**

**6.** Fill the petri dish to the halfway point with water. Set the wick of the filter paper into this water, as shown in the illustration, and wait for the chromatogram to develop.

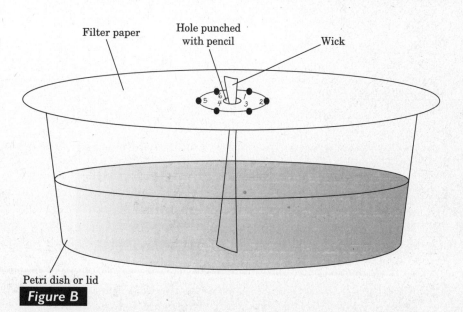

Filter paper     Hole punched with pencil     Wick

Petri dish or lid
**Figure B**

**7.** Repeat steps **5** and **6** with the second piece of filter paper. Instead of water, use isopropanol to fill the petri dish lid to the halfway point.

8. Allow the chromatograms to develop for approximately 15 min, or until the solvent is about 1 cm away from the outside edge of the paper. Remove each piece of filter paper from the petri dish and lid, and allow them to dry.

9. Record the colors that have separated on the chromatogram from each of the six different black inks in your data table. You may either describe the colors or use colored pencils to record this information.

10. Take the piece of the forged signature, and choose two segments that can be used to make chromatograms. (Hint: consider in which direction the ink will travel after the wick brings solvent onto the paper.) Cut out the two segments, leaving as much blank paper attached to each one as possible.

11. With one piece of the paper, make a chromatogram using the water in the petri dish. Do not allow the ink from the paper to come into direct contact with the solvent. Instead, use a wick as before. Unbend the paper clip, and use it to support the strip with the signature, as shown in Figure C. Then make a chromatogram using the isopropanol in the petri dish lid with the other piece of the paper.

Wick

Piece of forged signature

Unbent paper clip

Petri dish or lid

**Figure C**

12. After 15 min or when the solvent is about 1 cm away from the top of the paper, remove each chromatogram from the petri dish or lid, and allow them to dry.

13. Record the colors from the pieces of the signature and their arrangement in your data table.

## Cleanup and Disposal

14. The water may be poured down the sink. Chromatograms and other pieces of filter paper may be discarded in the trash can. The isopropanol solution should be placed in the waste disposal container designated by your teacher. Clean up your equipment and lab station. Thoroughly wash your hands after completing the lab session and cleanup.

## Analysis and Interpretation

1. **Organizing Ideas**
   Draw Lewis dot structures for water, $H_2O$, and isopropanol, $CH_3CHOHCH_3$.

2. **Analyzing Ideas**
   Analyze the bonding and structure of water and isopropanol molecules. What types of substances, polar or nonpolar, ionic or covalent, are most likely to dissolve in each one?

3. **Applying Ideas**
   Some components of ink are minimally attracted to the stationary phase and very soluble in the solvent. Where on the filter paper should these components be located in the final chromatogram?

4. **Relating Ideas**
   The diffusion of a solute in a solvent is somewhat similar to the diffusion of gases. Where on the filter paper do you expect the larger molecules to be located in the final chromatogram?

5. **Evaluating Methods**
   Explain why the labels numbering the pen spots were written in pencil. (Hint: recall what you know about the nature of the bonds and structure in the graphite that is found in pencils.)

6. **Predicting Outcomes**
   What would have happened to the chromatogram if the process had not been stopped after 15 min, but instead was allowed to proceed overnight?

7. **Predicting Outcomes**
   At Bryan High School, Armando was performing the chromatogram experiment. He thought it was going too slow and decided to work without a wick. He dipped the sample into the solvent so that the ink dot was even with the top of the water. But when he finished, the chromatogram was too faint to be seen. Explain why Armando's experiment failed.

## Conclusions

8. **Analyzing Conclusions**
   Are the properties of the component that traveled the furthest in the water chromatogram likely to be similar to the properties of the component that traveled the furthest in the isopropanol chromatogram? Explain your reasoning.

9. **Analyzing Conclusions**
   Which pen was used for the forged signature? Explain your reasoning, giving specific examples and providing quantitative data as possible (Hint: measure the distances that each component traveled to gather quantitative data.)

## Extensions

1. **Designing Experiments**
   How would you improve the efficiency of the separation? If you can think of a way to make the technique work better, ask your teacher to approve your plan, and run the procedure again.

## 2. Predicting Outcomes

Amino acids are organic chemicals that are monomers of proteins. They each have at least one amine group and one carboxylic acid group, each of which can carry charge in certain solutions. A scientist is analyzing a sample that contains a mixture of the four amino acids shown below. She can detect the presence of any amino acid by spraying it with an indicator called ninhydrin. She makes two chromatograms, one with water and another with isopropanol, using different portions of the sample. In what order would the amino acids be after each chromatogram was dried and sprayed with ninhydrin? (Hint: it may be helpful to refer to your answers to items **4** and **8**. Also consider what you know about solubility rules.)

Glycine

Valine

Aspartic acid

Phenylalanine

## 3. Applying Ideas

In a gel chromatography apparatus, a polymeric gel that has many pores, or openings, is packed into a long tube or column. A solvent is continually forced under low pressure through the tube. The gel serves as the stationary phase, and the solvent is the mobile phase. A sample of a mixture can be injected into the solvent just before it enters the column, and it will be separated into its component parts. Each component of the sample will take a different amount of time to pass through the column. If the solvent used is benzene, in what order will the plant pigments shown below pass through the column? (Hint: it may be helpful to refer to your answers to items **4** and **8**. Also consider what you know about solubility rules.)

Beta carotene (yellow/orange)

Delphinidin (blue/violet)

Betanidin (reddish)

**Chem**File

HOLT
ChemFile
LAB PROGRAM

# Paper Chromatography—Forensic Investigation

March 23, 1995

Marissa Bellinghausen
Director of Investigations
CheMystery Labs, Inc.
52 Fulton Street
Springfield, VA 22150

Dear Ms. Bellinghausen:

Late last night, someone broke into a display case at Frederick Douglass High School and stole a rare manuscript of a speech in Douglass's own writing that was part of a special exhibit.

While our detectives were examining the break-in area this morning, a hand-lettered ransom note asking for $100 000 was placed in the principal's mailbox.

We have been unable to pin down the time of the theft, but the ransom note was placed in the mailbox between 10:14, when the principal emptied it, and 10:25, when he noticed something else there.

Only four people had entered the hallway where mailbox is. Each of them agreed to submit to a search, but the missing manuscript was not found. However, each of the possible suspects was carrying a black pen that could have been used on the ransom note.

If we can obtain evidence proving which suspect wrote the note, we believe the perpetrator will surrender the priceless manuscript before it is damaged or sold.

Sincerely,

*Alyssa Hartman*

Alyssa Hartman
Captain, Eleventh Precinct
Springfield Police Department

# Memorandum

*Date:* March 24, 1995
*To:* Janey Markowitz
*From:* Marissa Bellinghausen

*CheMystery Labs, Inc.*
*52 Fulton Street*
*Springfield, VA 22150*

We need to act quickly. If we can demonstrate that we can provide accurate and prompt results, more work from law enforcement could come our way. Due to budget cutbacks, the police department can afford to spend only $75 000 on this job. Because of the number of tests that must be performed on the evidence, you will be given only a very small piece of the sample, so plan carefully!

Before you begin the job, Captain Hartman says the district attorney would like to make sure our technique will hold up in court. Prepare the following for me to send to them.

- detailed one-page plan for the procedure and all necessary data tables
- detailed list of the equipment and materials you will need, along with the individual and total costs

After I tell you that the district attorney is satisfied, you may begin your lab work. When your research team has completed the lab work, you should prepare a report in the form of a two-page letter to Captain Hartman. Remember that this document will become part of the police report and also could become evidence in a later trial. It should include all of the following.

- which pen was used to write the ransom note
- summary paragraph of how you solved this problem
- discussion of the certainty of your method
- detailed and organized analysis of the sample
- detailed and organized data table
- detailed invoice including costs for equipment and supplies as well as services rendered

## Required Precautions

 Always wear safety goggles and a lab apron to protect your eyes and clothing.

 Do not touch any chemicals. If you get a chemical on your skin or clothing, wash the chemical off at the sink while calling to your teacher.

 Isopropanol is flammable. No burners, flames, hot plates, or other heat sources should be present in the room while isopropanol is being used. Keep isopropanol in the hood.

 Always wash your hands thoroughly when finished.

## Spill Procedures/ Waste Disposal Methods

- In case of a spill, follow your teacher's instructions.
- Put excess isopropanol in the container designated by your teacher. Do not pour it down the sink.
- Any of the water may be poured down the sink.
- The chromatograms and filter paper wicks may be discarded in the trash can.

## References

Carlo Avena, the chemistry teacher, came in at 10:17 along with Johnny Pudstrom to discuss with the principal Johnny's latest excuse for not having his lab goggles, homework, or books. Mr. Avena, who owns Pen A, is said to be jealous that the biology teacher was named science department chair. Johnny has been in a number of scrapes this year, and the principal has put him on probation. Johnny owns Pen B, which was the only school supply he had with him.

Laney Femberton, a senior, brought in the collected attendance reports for the third period at 10:20. Laney is eager to find a way to make enough money so that she can quit school and her after-school job and devote more time to the beauty contest circuit. Laney owns Pen C.

The last possible suspect is the principal, Joshua Skinner. The superintendent has announced an audit of the school's finances in order to put to rest once and for all rumors of mismanagement. Skinner owns Pen D.

## MATERIALS for Forensic Investigation

### REQUIRED ITEMS
(You must include all of these in your budget.)

| | |
|---|---|
| Lab space/fume hood/utilities | 15 000/day |
| Standard disposal fee | 20 000/L of solution |

### REAGENTS AND ADDITIONAL EQUIPMENT
(Include in your budget only what you will need.)

| | |
|---|---|
| Isopropanol | 1 000/mL |
| 150 mL beaker | 1 000 |
| 250 mL beaker | 1 000 |
| 400 mL beaker | 2 000 |
| 250 mL Erlenmeyer flask | 1 000 |
| 100 mL graduated cylinder | 1 000 |
| Aluminum foil | 1 000/cm² |
| Balance | 5 000 |
| Cotton swab | 500 |
| Desiccator | 3 000 |
| Filter paper | 500/piece |
| Paper clips | 500/box |
| Petri dish and lid | 1 000 |
| Pipet or medicine dropper | 1 000 |
| Ruler | 500 |
| Stopwatch | 5 000 |
| Wax pencil | 500 |

\* No refunds on returned chemicals or unused equipment.

### FINES

| | |
|---|---|
| OSHA safety violation | 2 000/incident |

HOLT
ChemFile
LAB PROGRAM

# Testing for Dissolved Oxygen

## Situation

The company you work for has been hired as an expert witness in a lawsuit. The local chapter of Bass Anglers Unlimited has been disturbed by recent declines in the population of bass in Pulaski Lake. There have been several fish kills, in which large numbers of fish die at the same time and float to the surface of the lake, creating a terrible smell and fouling the water. The fishermen claim that the fish kills began shortly after the R. C. Throckmorton Power Plant came on line, and they are seeking a court order to shut down the plant. The machinery in the power plant uses water from the lake as a coolant and then returns the water to the lake.

The fishermen say that something in the returned water is killing the fish. The utility company operating the power plant points out that they use a closed system that prevents the water from coming into direct contact with the machinery in the plant. They say that the water is just as pure when it comes out; it's just a little warmer. The court has asked you to investigate whether there is a scientific basis for the fishermen's claim.

## Background

Fish rely on the oxygen dissolved in water to live. The water passes through their gills, which remove the oxygen. The less oxygen in the water, the harder the gills have to work to get enough oxygen to keep the fish alive. The normal lake temperature is between 15°C and 17°C. Another witness, a biology professor specializing in ichthyology (the study of fish) has testified already that if the oxygen content of the water dips below 90% of the value at these temperatures, the fish will suffer long-term damage or death. The temperature of the water returned to the lake from the power plant is 28°C. To measure the change in dissolved oxygen content over this temperature range, you can use tap water, which has a mineral content similar to lake water.

## Problem

In order to evaluate these claims, you will need to do the following.
- Prepare water samples at several different temperatures.
- Measure the dissolved oxygen content in ppm for each one.
- Graph the relationship between the solubility of dissolved oxygen and temperature.
- Extrapolate to determine the solubility of dissolved oxygen at 16°C and 28°C.
- Compare the solubilities to determine if there is enough difference to cause damage to the fish.

## Objectives

***Measure*** the concentration of oxygen in a sample of water.

***Graph*** the relationship between the concentration of a gas and temperature.

***Infer*** a general rule of thumb for gas solubilities and temperature.

***Relate*** changes in gas solubility to a fish kill.

## Materials

- Ice
- 600 mL beaker
- Beaker tongs
- Hot mitt
- Jars with screw-on lids, 4

**Bunsen burner option**

- Bunsen burner and related equipment
- Ring stand
- Ring clamp
- Wire gauze with ceramic center

**Hot plate option**

- Hot plate

**Probe option**

- Thermistor probe

**Alternative option**

- Dissolved oxygen test kit or dissolved oxygen meter
- Thermometer, non-mercury

## Safety

 Always wear safety goggles and a lab apron to protect your eyes and clothing. If you get a chemical in your eyes, immediately flush the chemical out at the eyewash station while calling to your teacher. Know the locations of the emergency lab shower and eyewash station and the procedure for using them.

 When you use a Bunsen burner, confine any long hair and loose clothing. Do not heat glassware that is broken, chipped, or cracked. Use tongs or a hot mitt to handle heated glassware and other equipment because hot glassware does not always look hot. If your clothing catches on fire, WALK to the emergency lab shower, and use it to put out the fire.

 Always clean up the lab and all equipment after use, and dispose of substances according to proper disposal methods. Wash your hands thoroughly before you leave the lab after all lab work is finished.

## Preparation

1. **Organizing Data**
   Prepare a data table in your lab notebook with two columns and five rows. In the first row, label the two columns *Temp. (°C)* and *Dissolved O₂ (ppm)*.

2. Label four jars *Ice water*, *Room temp.*, *50°C*, and *100°C*.

## Technique

### Sample Preparation

3. Add approximately 50 g of ice to 100 mL of tap water in the *Ice water* jar. Let the ice-water mixture stand for 5 min. Measure the temperature with a thermometer or a thermistor probe to the nearest 0.1°C. The temperature should be near 4°C. Record the temperature in your data table. Add more ice until the jar is filled to the rim. Screw on the top and place the sample in an ice chest or refrigerator overnight.

4. Fill the *Room temp.* jar with tap water. Do not seal the jar. Let it sit out open overnight where it will not be disturbed. You will measure the temperature of this sample tomorrow.

5. If you are using a Bunsen burner, set up the ring stand, ring, and wire gauze over the burner so that they will hold a beaker. If you are using a hot plate, continue with step **6.**

6. Pour approximately 450 mL of tap water into a 600 mL beaker, and gently heat to about 50°C. Maintain 50°C as closely as possible for 5 min. Measure the temperature of the water with a thermometer or a thermistor probe to the nearest 0.1°C. Record the temperature in your data table. Using beaker tongs to hold the hot beaker, gently fill the jar labeled *50°C* with the warm water. Fill the container to the rim. Screw on the top, and store in a safe place.

7. Heat the remaining water to boiling (approximately 100°C). Allow the water to boil for approximately 30 min. Measure the temperature of the water with a thermometer or a thermistor probe to the nearest 0.1°C. Record the temperature in your data table. Using beaker tongs to hold the hot beaker, gently fill the jar labeled *100°C* with the boiling water. Fill the container to the rim. Using a hot mitt, screw on the top and store in a safe place.

### Sample Testing

8. On the following day, retrieve your samples.

9. Measure the temperature of the water in the *Room temp.* jar with a thermometer or a thermistor probe to the nearest 0.1°C. Be sure to disturb the water as little as possible. Record the temperature in your data table.

10. If you are using a dissolved oxygen probe or meter, it may be necessary to calibrate it using a standardized solution. Ask your teacher for instructions. If your teacher indicates that this is unnecessary, continue with step **11**.

11. Disturbing the water as little as possible, measure the dissolved oxygen content of each sample, and record the value to the nearest 0.1 ppm in your data table. If you are using a dissolved oxygen test kit, you will be comparing the colors of standard solutions to the colors of your tested solutions. Estimate the measured concentrations to the nearest 0.5 ppm.

### Disposal

12. If you used chemical test kits, dispose of the ampuls and reagents in the containers designated by your teacher. Rinse the samples down the drain. Clean all equipment.

## Analysis and Interpretation

1. **Organizing Data**
   Make a graph of your data, with temperature (in °C) plotted on the x-axis and $O_2$ concentration (in ppm) plotted on the y-axis. Draw a straight line that best fits the data.

2. **Interpreting Graphics**
   Refer to the graph from item 1, and explain in your own words the relationship between temperature and the solubility of oxygen in water.

3. **Analyzing Methods**
   Why was it important to disturb the water as little as was possible when measuring the dissolved oxygen content?

4. **Applying Models**
   Often, when water is heated, small bubbles appear long before boiling begins. Explain why these bubbles form. (Hint: use the concept of solubility in your explanation.)

## Conclusions

5. **Interpreting Graphics**
   Using the graph from item 1, extrapolate to determine the concentration of $O_2$ in the water at 16°C and 28°C.

6. **Interpreting Graphics**

Using the format shown below, determine the equation for the line you drew on the graph in item **1**. (Hint: if you have a graphics calculator, use the $\boxed{STAT}$ mode to enter your data and make a linear regression equation using the LinReg function from the STAT menu.)

$$O_2 \text{ conc.} = m(\text{temp.}) + b$$

7. **Analyzing Conclusions**

Calculate the expected concentration of $O_2$ in the water at 16°C and 28°C, using the equation from item **6**.

8. **Evaluating Conclusions**

Using the values from item **5** as the accepted values, calculate the percent error for the values calculated in item **7**.

9. **Evaluating Conclusions**

Earlier in the trial, the ichthyologist testified that fish receiving less than 90% of the oxygen available at 16°C would suffer long-term damage. From item **5** and item **7**, determine what percentage of the oxygen available at 16°C is available at 28°C. Is it likely that the water is damaging to the fish?

## Extensions

1. **Evaluating Conclusions**

During cross-examination, the attorney for the power plant suggests that because you tested tap water instead of water taken directly from the power plant, your solubility results are irrelevant, and the power plant's water could have even more oxygen in it than the original cold water. Is the attorney correct? Why or why not?

2. **Evaluating Conclusions**

Later in the cross-examination, the attorney for the power plant asks whether your results establish conclusively that the fish that were killed died because of the warmer water and for no other reasons. What do you say?

3. **Designing Experiments**

Assuming that you had unlimited laboratory resources and access to all of the lake, what other tests would you perform to be more certain of your results?

4. **Evaluating Methods**

The judge has asked for the opinions of all expert witnesses about a proposed settlement. The power plant proposes to insert a device to bubble oxygen through the warm water as it is released into the lake. Will this solve the problem? Explain why or why not, using the principles of solubility.

5. **Relating Ideas**

The power plant supplies 10 000 households with their electricity. The judge has asked for possible solutions that will keep the power plant working and prevent further damage to the fish. What do you suggest?

HOLT
ChemFile
LAB PROGRAM

# Solubility of Ammonia

## Situation

Sparkle-Clean Industries has contacted you about one of their products, a household window cleaner with ammonia, $NH_3$. Sparkle-Clean did not have a product for this market until they bought out a competitor. The competitor purchased gaseous $NH_3$ from a local gas products supplier. Then the $NH_3$ was dissolved in water at room temperature as the first and most difficult step in the manufacturing process.

Sparkle-Clean's manufacturing equipment has several features that the competitor did not have. One piece of equipment can maintain liquids at temperatures from 0°C to 80°C. Perhaps changing the temperature could improve the competitor's procedure so that more $NH_3$ would dissolve in the water. However, the equipment would need expensive adjustments to keep the hazardous $NH_3$ gas under control. To determine whether the savings will justify the initial expense, they want you to model the process on a small scale, determine the best temperature, and estimate how much $NH_3$ dissolves at that temperature.

## Background

Most gases dissolve better in water at cooler temperatures, provided that the pressure remains the same. This is the opposite of the pattern observed for most solids. One example is a soda bottle, the warmer the soda, the flatter it will taste because the dissolved $CO_2$ that provides the "fizz" will bubble out of solution. You must test the $NH_3$ to see if it follows the same pattern. Measuring how long it takes for a given amount of $NH_3$ to dissolve can also provide a rough measure of how much more efficient the process will be at a different temperature.

## Problem

To determine the optimum temperature for this procedure, you must do the following.
- Prepare some $NH_3$ gas by heating a solution of aqueous $NH_3$.
- Test the $NH_3$ by dissolving it in water at various temperatures.
- Compare efficiencies of the dissolving process for the $NH_3$ in the different water samples by measuring the time for it to dissolve completely.

## Objectives

***Prepare*** a small amount of ammonia gas in a container.

***Calculate*** amount of ammonia present, given volume, temperature, and pressure.

***Test*** how well the ammonia dissolves in water at different temperatures.

***Relate*** changes in volume of a gas to the removal of gas particles as they dissolve in a solution.

## Materials

- NH₃(aq), concentrated
- Distilled water
- Ice
- 250 mL beakers, 6
- 400 mL beaker
- Pipet, thin-stem
- Pipet, wide-stem
- Ruler, metric
- Scissors
- Stopwatch or clock with second hand
- Thermometer, non-mercury

**Bunsen burner option**

- Bunsen burner and related equipment
- Ring stand
- Ring clamp
- Wire gauze with ceramic center

**Hot plate option**

- Hot plate

**Probe option**

- Thermistor probe

—4.0 cm

←— cut here

**Figure A**

## Safety

**Always wear safety goggles and a lab apron to protect your eyes and clothing. If you get a chemical in your eyes, immediately flush the chemical out at the eyewash station while calling to your teacher. Know the locations of the emergency lab shower and eyewash station and the procedure for using them.**

**Because the procedure involves poisonous ammonia fumes, the procedure must take place in a properly functioning fume hood. Make sure that the fume hood is turned on before starting the reaction.**

**Do not touch any chemicals. If you get a chemical on your skin or clothing, wash the chemical off at the sink while calling to your teacher. Make sure you carefully read the labels and follow the directions on all containers of chemicals that you use. Never return leftovers to their original containers; take only small amounts to avoid wasting supplies.**

**Ammonia is a base. Call your teacher in the event of a spill of the NH₃ solution. Acid or base spills should be cleaned up promptly, according to your teacher's instructions.**

**When you use a Bunsen burner, confine any long hair and loose clothing. Do not heat glassware that is broken, chipped, or cracked. Use tongs or a hot mitt to handle heated glassware or other equipment that has been heated because hot equipment does not always look hot. If your clothing catches on fire, WALK to the emergency lab shower and use it to put out the fire.**

**Always clean up the lab and all equipment after use, and dispose of substances according to proper disposal methods. Wash your hands thoroughly before you leave the lab after all lab work is finished.**

## Preparation

1. **Organizing Data**
   Prepare a data table in your lab notebook with five rows and two columns. Label the columns in the first row *Temp. (°C)* and *Time (s)*. You will also need room beneath the data table to record the total volume of the ammonia sample.

2. Cut off the tip of the wide-stem pipet at an angle so that the stem is as close to 4 cm long as possible, as shown in Figure A. This will be the collecting pipet in which the ammonia gas is collected.

3. If you were given the correct type of bulb and followed the directions in step **2** exactly, the volume of the bulb should be about 5.0 mL. If you used different equipment, ask your teacher how to measure the volume of the bulb. You will need a 10 mL or 25 mL graduated cylinder to measure the bulb's volume.

4. One of the 250 mL beakers will already be labeled *Conc. NH₃(aq)* by your teacher. It should already contain concentrated ammonia solution. Label the remaining five beakers *Holder, 80°C, 50°C, Room temp.,* and *0°C.*

## Technique

5. While working in the hood, prepare a warm-water bath with about 350 mL of water in the 400 mL beaker using a Bunsen burner and a ring stand or a hot plate. The temperature of the water should be about 80°C.

6. Pour 200 mL of the hot water into the *80°C* beaker. Keep it under the hood and not far from the warm-water bath. Refill the warm-water bath.

7. Working in the hood, fill the narrow-stem pipet about halfway with concentrated aqueous ammonia from the *Conc. NH₃(aq)* beaker. This pipet is the gas generating pipet. Turn it upside down (bulb end down). Keep it under the hood at all times.

8. Loosely cover the tip of the generating pipet with the bulb of the collecting pipet, as shown in Figure B. Be sure that the air in the collecting pipet can escape as the NH₃ enters and displaces the air.

9. Hold the pipet assembly with the bulb of the generating pipet dipped into the warm-water bath under the hood so that the aqueous NH₃ warms. As the solution warms, NH₃ gas is driven off into the collecting bulb. You will notice the aqueous NH₃ solution bubbling. Squeeze the collecting pipet several times to expel all of the air and allow the NH₃ gas to completely fill the collecting pipet.

10. Measure the temperature of the water in the *80°C* beaker with a thermometer or a thermistor probe to the closest 0.1°C. Record the temperature in your data table.

11. After 2–3 min, remove the generating pipet and collecting pipet from the warm-water bath. Place the generating pipet with the bulb end down into the *Holder* beaker.

12. Using a stopwatch or having your partner keep an eye on a clock with a second hand, place the tip of the collecting pipet into the water in the *80°C* beaker, as shown in Figure C. Record your observations in the data table. Measure and record how many seconds it takes until the gas is completely dissolved and the bulb is filled with water.

13. Pour about 200 mL of the water from the warm-water bath into the *50°C* beaker. Refill the warm-water bath.

14. Using the procedure from steps **7–9,** work under the hood to collect another bulb of NH₃.

15. Add ice to the water until the temperature of the water in the *50°C* beaker is close to 50°C. Empty some of the water until about 200 mL remains.

16. Measure the temperature of the water in the *50°C* beaker with a thermistor probe or a thermometer, and record it in the data table. Return this beaker to the hood.

Collecting pipet

Generating pipet

NH₃*(aq)*

**Figure B**

**Figure C**

**88** EXPLORATION C14

**Chem**File

17. Using a stopwatch or having your partner keep an eye on a clock with a second hand, place the tip of the collecting pipet into the water in the *50°C* beaker. Record your observations in the data table. Measure and record how many seconds it takes until the gas is completely dissolved.

18. Repeat steps **14–17,** working under the hood with 200 mL of water at room temperature in the *Room temp.* beaker and with 200 mL of an ice-water mixture in the *0°C* beaker.

## Disposal

19. The solutions of aqueous ammonia should be left in a disposal container in the hood. Clean all other equipment after use.

## Analysis and Interpretation

1. *Analyzing Methods*
   How can you be certain that the air will be displaced out of the bulb instead of the $NH_3$? (Hint: air is mostly $O_2$ and $N_2$. Of the three gases, which two are likely to be the densest?)

2. *Analyzing Methods*
   Use your own words to explain how and why the $NH_3$ gas was generated when the $NH_3$ solution was heated.

3. *Analyzing Data*
   Using the ideal gas law, calculate the number of moles of $NH_3$ present from the initial volume in the bulb and pipet tip before it is tested in the water. (Hint: you measured the temperature of room temperature water, and the pressure is about 1.0 atm in the lab.)

4. *Analyzing Data*
   Using the periodic table and your answer from item **3,** calculate the mass of $NH_3$ present in the bulb at the beginning.

5. *Analyzing Information*
   Explain why the water was pushed into the collecting bulb. (Hint: what happened to some of the $NH_3$ when it came into contact with water?)

## Conclusions

6. *Evaluating Conclusions*
   Which temperature would be the most effective for Sparkle-Clean to use in their new process? Explain why.

7. *Evaluating Conclusions*
   If the process at room temperature dissolves 500. g of ammonia gas in 15 min, approximate how long it would take to dissolve 500. g of ammonia gas at the temperature from item **6.**

8. *Evaluating Conclusions*
   Can you tell from your data how much more $NH_3$ will dissolve at the most effective temperature?

9. *Applying Ideas*
   Using information on solubility and polarity explain why this procedure works for dissolving $NH_3$ in water, but not for dissolving $CO_2$ to make a soft drink.

# Extensions

## I. Applying Ideas

Actually, the $NH_3$ that was collected was not pure. Because it was collected over water, some of the gas in the bulb was water vapor. Using the vapor pressure value for water that matches the conditions in your lab, recalculate the amount of $NH_3$ in the bulb before each of the trials. (Hint: 1.0 atm = 101.325 kPa)

| Temperature (°C) | Partial Pressure of $H_2O$ (kPa) |
|---|---|
| 16 | 1.82 |
| 17 | 1.94 |
| 18 | 2.06 |
| 19 | 2.19 |
| 20 | 2.34 |
| 21 | 2.49 |
| 22 | 2.64 |
| 23 | 2.81 |
| 24 | 2.98 |
| 25 | 3.17 |
| 26 | 3.36 |
| 27 | 3.56 |
| 28 | 3.78 |

## 2. Predicting Outcome

According to Henry's law, the solubility of a gas is directly proportional to the partial pressure of the gas over the surface of the water. Sparkle-Clean can run the procedure at pressures of 1.0, 1.5, and 2.0 atm. Which should they pick to maximize the amount of $NH_3$ dissolving in the water? How much $NH_3$ should dissolve at this pressure?

## 3. Designing Experiments

What possible sources of error can you identify with this procedure? If you can think of ways to eliminate them, ask your teacher to approve your plan, and run the procedure again.

## 4. Research and Writing

Studies indicate that much of the $CO_2$ that is generated by living things and by the combustion of fossil fuels is dissolving in ocean water. Assuming this is true, explain how the $CO_2$ levels in the atmosphere would be affected if the environmental models that predict a period of global warming are correct.

**Chem**File

# Solubility Product Constant

## Situation

Juliette Brand Foods is planning to add dill pickles to their product line. Pickled foods require soaking or packing in a concentrated solution of sodium chloride called brine. The company intends to make their own brine in large quantities. To help them set up their process, they have asked your company to determine the mass of NaCl that should be added to water in a 4000.0 L vat at room temperature so that the solution will have the highest possible concentration with no NaCl remaining undissolved. Juliette Brand Foods is also considering the use of sodium citrate, $Na_3C_6H_5O_7$, as a flavoring agent. They need you to determine what effects, if any, on the solubility of NaCl the addition of $Na_3C_6H_5O_7$ will have.

## Objectives

***Prepare*** a saturated solution of sodium chloride.

***Determine*** the mass of sodium chloride dissolved in a volume of saturated solution.

***Calculate*** the solubility of sodium chloride.

***Determine*** the solubility product constant of sodium chloride.

***Demonstrate*** the effect of the addition of a common ion to a saturated solution.

## Background

A saturated solution contains the maximum amount of solute that can dissolve when added to a solvent at a given temperature. Saturated solutions usually contain an excess of undissolved solid so that the ions in solution are at equilibrium with the undissolved solid. The equilibrium constant that describes this situation is called the solubility product constant, $K_{sp}$. For salts composed of ions with single positive or negative charges, the $K_{sp}$ is the product of the concentrations of undissolved ions. If either of the ion concentrations is increased after a saturated solution is formed, the value of the $K_{sp}$ will be exceeded, and precipitation will occur until equilibrium is restored. This phenomenon is called the "common-ion effect."

## Problem

To answer your client's questions you need to do the following.
- Make a saturated solution of NaCl at room temperature.
- Calculate the concentration in moles per liter of each ion in the solution.
- Write the equilibrium expression and calculate the $K_{sp}$ of NaCl at room temperature.
- Observe the effect of adding $Na_3C_6H_5O_7$ to a saturated solution of NaCl.
- Use your data to calculate the mass of NaCl necessary to fill the pickling vat with 4000.0 L of saturated solution.

## Safety

 Always wear safety goggles and a lab apron to protect your eyes and clothing. If you get a chemical in your eyes, immediately flush the chemical out at the eyewash station while calling to your teacher. Know the locations of the emergency lab shower and eyewash station and the procedure for using them.

 Do not touch any chemicals. If you get a chemical on your skin or clothing, wash the chemical off at the sink while calling to your teacher. Make sure you carefully read the labels and follow the directions on all containers of chemicals that you use. Do not taste any chemicals or items used in the laboratory. Never return leftovers to their original containers; take only small amounts to avoid wasting supplies.

 When you use a Bunsen burner, confine any long hair and loose clothing. Do not heat glassware that is broken, chipped, or cracked. Use tongs or a hot mitt to handle heated glassware or other equipment that has been heated because hot equipment does not always look hot. If your clothing catches on fire, WALK to the emergency lab shower, and use it to put out the fire.

 Always clean up the lab and all equipment after use, and dispose of substances according to proper disposal methods. Wash your hands thoroughly before you leave the lab after all lab work is finished.

## Materials

- Distilled water
- NaCl, 15 g
- $Na_3C_6H_5O_7 \cdot 2H_2O$, 0.50 g
- 150 mL beakers, 2
- 10 mL graduated cylinder
- 25 mm × 100 mm test tube
- Balance
- Crucible tongs
- Evaporating dish
- Glass stirring rod
- Spatula
- Weighing paper

**Bunsen Burner option**

- Bunsen burner and related equipment
- Ring stand and ring
- Wire gauze with ceramic center, 2

**Hot Plate option**

- Hot plate

## Preparation

1. **Organizing Data**
   Prepare a table for data and observations. You will need space to record the mass of the empty evaporating dish, the volume of saturated NaCl solution, and the mass of the evaporating dish and NaCl. You will also need a space to record the mass of $Na_3C_6H_5O_7 \cdot 2H_2O$ used. Be certain there is additional room for recording your observations from step **11**.

2. Make sure that your equipment and tongs are very clean for this work so that you will get the best possible results. Remember that you will need to cool the evaporating dish before measuring its mass. *Never put a hot evaporating dish on a balance; it may damage the balance, and the currents of hot air will cause an error in the mass measurement.*

## Technique

3. Prepare a saturated solution of NaCl by adding 15 g of the salt to 25 mL of water in a beaker. Stir constantly until it appears that no more salt will dissolve. There should be some solid on the bottom of the beaker. Decant the saturated solution into another beaker, making certain that no solid is transferred. Pour the excess solid and the remaining solution down the drain with plenty of water.

Evaporating dish

Wire gauze
with ceramic center

Low
flame

Students may hold
burner here when
heating dish gently

**Figure A**

4. Using tongs, place the evaporating dish on the balance and measure its mass to the nearest 0.01 g. Record the mass in your data table.

5. Using a graduated cylinder, measure approximately 10.0 mL of the salt solution. Record the volume to the nearest 0.1 mL in your data table. Pour this sample of saturated NaCl solution into the evaporating dish.

6. Set up the ring stand assembly and Bunsen burner as shown in Figure A. The ring and wire gauze should be positioned several centimeters above the burner flame. Place the evaporating dish on the wire gauze, and light the burner. Heat the NaCl solution slowly to avoid spattering. Continue heating until all of the water has evaporated and the remaining crystals are dry.

7. When the remaining crystals appear dry, adjust the burner to produce a very low flame and continue heating for another 10 min. Hold the burner by the base, and gently wave the flame under the evaporating dish, being careful not to heat the material so rapidly that it spatters.

8. Using tongs, place the hot evaporating dish on a piece of wire gauze at the base of the ring stand and allow it to cool. Then determine the mass of the evaporating dish and the solid residue to the nearest 0.01 g. Record the mass in your data table.

9. Measure 5.0 mL of saturated NaCl solution, and pour it into a test tube.

10. Using a spatula, measure about 0.50 g $Na_3C_6H_5O_7 \cdot 2H_2O$ on the balance. Record the exact mass in your data table.

11. Add the $Na_3C_6H_5O_7 \cdot 2H_2O$ to the test tube and stir. Record your observations.

## Cleanup and Disposal

12. All of the solutions in this lab may be disposed of by washing them down the drain. Place any excess NaCl or $Na_3C_6H_5O_7 \cdot 2H_2O$ in the disposal container designated by your teacher.

## Analysis and Interpretation

1. **Organizing Data**
   Determine the mass of dry NaCl in the evaporating dish by subtracting the mass of the evaporating dish from the mass of the evaporating dish with the NaCl.

2. **Organizing Data**
   Use the periodic table to determine the molar mass of NaCl. Then calculate the number of moles of NaCl in the evaporating dish.

3. **Analyzing Data**
   Calculate the molar concentration of the 10.0 mL of NaCl solution that you evaporated.

4. **Organizing Ideas**
   Write the chemical equation for dissolving NaCl to form the saturated solution.

5. **Organizing Data**
   Use the concentration from item **3** and the equation from item **4** to determine the concentrations of $Na^+$ and $Cl^-$ ions in the saturated solution.

6. **Applying Ideas**
   Write the equilibrium expression for the equation in item **4**.

7. **Analyzing Data**
   Use your answers to items **5** and **6** to calculate the numerical value of the $K_{sp}$ for NaCl.

8. **Organizing Data**
   How many moles of $Na_3C_6H_5O_7 \cdot 2H_2O$ were added to the solution in step **11**?

## Conclusions

9. **Applying Conclusions**
   Use your results from item **3** to determine the mass of NaCl needed to make 4000.0 L of saturated solution.

10. **Analyzing Information**
    Using the equilibrium expression from item **6,** explain your observations from step **11,** when $Na_3C_6H_5O_7 \cdot 2H_2O$ was added to the NaCl solution.

11. **Predicting Outcomes**
    The development team at Juliette foods is also considering adding the following flavoring or preservative additives to the brine: acetic acid, aluminum sulfate, calcium chloride, sodium bicarbonate, and sodium nitrate. Indicate which ones will cause a change in the NaCl equilibrium concentration.

## Extensions

1. **Applying Conclusions**
   Suppose your client diluted the saturated NaCl brine to one half of its molarity. Would it be possible then to add 50.0 g of $CaCl_2$ to every liter of the diluted solution without causing precipitation? Show calculations to support your answer.

2. **Applying Ideas**
   Although iron(III) nitrate and iron(III) acetate have very large solubility product constants, farmers do not find these compounds useful as additives for iron-deficient soils. Explain why this is true.

3. **Applying Ideas**
   Assuming that the value for $K_{sp}$ of NaCl stays the same and that all of the $Na_3C_6H_5O_7 \cdot 2H_2O$ dissolves, you can calculate the new equilibrium concentration for $Na^+$ and $Cl^-$ using the following steps.
   • Calculate the total possible concentration of $Na^+$ ions.
   • Assume that $x$ mol/L of NaCl precipitates.

- Rewrite the equilibrium expression with $[Na^+]$ equal to the total possible concentration minus $x$ and $[Cl^-]$ equal to the initial concentration minus $x$.
- Algebraically rearrange the equation until there is a zero on one side of the equation.
- Apply the quadratic formula to the equation from the previous step to determine the value of $x$, the equilibrium concentration of NaCl. For any equation with the form given below, $x$ has the values shown.

$$ax^2 + bx + c = 0$$

$$x = \frac{-b \pm \sqrt{b^2 - 4ac}}{2a}$$

4. **Research and Communication**
   The hollowing out of limestone caves and the growth of stalactites and stalagmites are environmental examples of solubility equilibria at work. Investigate these processes in terms of equilibrium, and prepare a paper explaining the formation of caves.

5. **Applying Ideas**
   The method for producing sweet pickles is essentially the same as that for making dill pickles, except that the cucumbers are soaked in a saturated sucrose solution instead of a NaCl solution. (Incidentally, sweet pickles were invented when the person doing the pickling mistakenly put sugar into the water instead of salt.) Depending upon the success of their new line of dill pickles, Juliette Brand Foods is planning to expand production to include sweet pickles. Based on your knowledge of how sugar dissolves in water, would you be able to experimentally determine a solubility-product constant for sucrose? Does this affect the precipitation of sugar in the presence of a common ion? Explain your answers.

6. **Predicting Outcomes**
   Because they are just beginning to enter this market, Juliette Brand Foods does not want to have to spend more capital than necessary on equipment. Included in this would be a temperature control for the brine vats, so that the solution remains at a constant temperature at all times. Does the solubility of NaCl vary sufficiently over a range of temperatures to make such a temperature control device necessary? The dependence of solubility on temperature for several solutes is shown in the textbook. Use this information to infer how the solubility product constant for NaCl varies with temperature.

7. **Relating Ideas**
   Often it is convenient to express the solubility product constant as $pK_{sp}$, which is defined similarly to pH.

$$pK_{sp} = -\log K_{sp}$$

   From this definition, what is the value of $K_{sp}$ for a $pK_{sp}$ value of 2.5? 25.7? 132?

# Solubility Product Constant—Algae Blooms

April 18, 1995

Ms. Sandra Fernandez
Director of Development
CheMystery Labs, Inc.
52 Fulton Street
Springfield, VA 22150

Dear Ms. Fernandez:

Recently, our region has experienced above-average temperatures and rainfall. These factors have increased runoff into local waterways, causing abnormal algae blooms in lakes and ponds. Our studies show that algae can be controlled with 0.0500 M solutions of copper(II) ions, so we are considering treating affected ponds with copper(II) sulfate or copper(II) chloride.

We would like to apply them to the ponds and lakes in the form of concentrated solutions. In this way, they will mix thoroughly with the lake or pond water much more quickly than if we added the solid compounds.

We are requesting bids for several comparative studies of the solubility properties of these two compounds. We want each contractor to recommend one compound or an optimum combination of both for our use. Please base any calculations of amounts necessary on a 3.90 × 10⁶ L pond and submit experimental data to support your conclusions.

Sincerely,

*Kathleen Farros-Hoeppner*

Kathleen Farros-Hoeppner
Assistant Director, Research
State Department of Fish and Game

# Memorandum
*Inc.*

*CheMystery Labs,*

*52 Fulton Street*
*Springfield, VA 22150*

*Date*: April 19, 1995
*To*: Gary Vasileyev
*From*: Sandra Fernandez

I think the best way to proceed is to <u>measure the concentrations of saturated solutions</u> of each compound. Make a 25 mL saturated solution of each one and <u>calculate $K_{sp}$</u> so that our results can be compared quickly to those of other firms.

If we land one of these contracts, it will be our <u>first project for a state agency</u>, so we want to make a favorable first impression. Send Ms. Farros-Hoeppner a bid that includes the following.
- detailed <u>one-page plan</u> for the procedure
- examples of all necessary <u>data tables</u>
- list of the <u>materials</u> you need with the <u>total cost</u>

Get started <u>as soon as your plan is approved</u>. When the work is complete, prepare a report in the form of a <u>two-page letter</u> to Kathleen Farros-Hoeppner. The letter must include the following items.
- your <u>recommendation</u> of which compound to kill the algae (be sure to consider costs of each one)
- <u>volumes of each saturated solution</u> needed to achieve a concentration of 0.05 M Cu(II) ions in the pond
- <u>mass</u> of each compound needed to make the solutions
- <u>equilibrium molar concentrations</u> of saturated solutions
- <u>experimentally measured $K_{sp}$</u> of copper(II) sulfate and copper(II) chloride based on experimental data
- brief description of the <u>procedure</u>
- <u>detailed and organized data table</u> and calculations
- <u>detailed invoice</u> for materials and services

## Required Precautions

 Always wear safety goggles and a lab apron to protect your eyes and clothing.

 When you use a Bunsen burner, tie back loose clothing and hair. Do not heat glassware that is broken, chipped, or cracked. Use tongs whenever handling glassware or other equipment that has been heated.

 Do not touch any chemicals. If you get a chemical on your skin or clothing, wash the chemical off at the sink while calling to your teacher.

 Always wash your hands thoroughly when finished.

## Spill Procedures/ Waste Disposal Methods

- In case of spills, follow your teacher's instructions.
- Dispose of the solid wastes in the waste container designated by your teacher.
- Dispose of the liquid wastes in the waste container designated by your teacher.

## MATERIALS for State Department of Fish and Game

### REQUIRED ITEMS
*( You must include all of these in your budget.)*

| | |
|---|---|
| Lab space/fume hood/utilities | 15 000/day |
| Standard disposal fee | 2 000/g of product |
| Balance | 5 000 |
| Crucible tongs | 2 000 |
| Evaporating dish | 1 000 |

### REAGENTS AND ADDITIONAL EQUIPMENT
*( Include in your budget only what you'll need.)*

| | |
|---|---|
| $CuCl_2 \cdot 2H_2O$ | 200/g |
| $CuSO_4 \cdot 5H_2O$ | 100/g |
| 150 mL beaker | 1 000 |
| 400 mL beaker | 2 000 |
| 250 mL flask | 1 000 |
| 25 mL graduated cylinder | 1 000 |
| 100 mL graduated cylinder | 1 000 |
| Filter paper | 500/piece |
| Glass stirring rod | 1 000 |
| Bunsen burner/related equipment | 10 000 |
| Hot plate | 8 000 |
| Litmus paper | 1 000/piece |
| Ring stand/ring/wire gauze | 2 000 |
| Ruler | 500 |
| Spatula | 500 |
| Test tube (large) | 1 000 |
| Test tube (small) | 500 |
| Wash bottle | 500 |

\* No refunds on returned chemicals or unused equipment.

### FINES

| | |
|---|---|
| OSHA safety violation | 2 000/incident |

## References

Reread the sections of your textbook that discuss information about saturated solutions and molarity. Review the background information about equilibrium systems, and the equilibrium constant.

   Hint: because these compounds can form hydrates, $CuCl_2 \cdot 2H_2O$ and $CuSO_4 \cdot 5H_2O$, they should be heated thoroughly to produce the anhydrous compounds after the solvent is evaporated. $CuSO_4$ should be gray or white (not blue or yellow), and $CuCl_2$ should be light brown.

# Freezing-Point Depression—Testing De-icing Chemicals

## Objectives

***Measure*** freezing-point depression for three solutions of unknown solutes.

***Calculate*** the molality of the particles in the solutions from freezing-point data.

***Demonstrate*** how ionic and non-ionic solutes differ in their effect on the freezing point of the solvent.

***Determine*** the best solute for melting ice on roads.

## Situation

The Department of Transportation (DOT) in your town is working to minimize hazardous driving conditions during the winter. They are considering adding a soluble substance to the sand that they spread on roads to prevent skidding. The purpose of the soluble material is to lower the freezing point so that the ice will melt. DOT has asked your firm to investigate several potential de-icing agents. They need quantitative freezing-point depression data and your recommendation of which solute is the most effective agent.

## Background

Review freezing-point depression, one of the colligative properties of solutions discussed in your textbook. Colligative properties depend on the number of particles present in a solution. Because ionic solutes dissociate into ions, they have a greater effect on freezing point than molecular solids of the same molal concentration. For example, the freezing point of water is lowered by 1.86°C with the addition of any nonvolatile molecular solute at a concentration of 1 molal. However, a 1 molal sodium chloride solution contains a 2 molal concentration of ions. Thus, the freezing-point depression for NaCl is 3.72°C, double that of a molecular solute. The relationship is given by the following equation.

$$\Delta t_f = K_f \times m \times n$$

$\Delta t_f$ is the freezing-point depression of the solution, $K_f$ is the molal freezing-point constant ($-1.86°C/m$ for aqueous solutions), $m$ is the molality of the solution, and $n$ is the number of particles formed from the dissociation of each formula unit.

## Problem

Other workers have prepared and frozen solutions containing 100.0 g of each de-icing chemical dissolved in 1 kg of water. To make a recommendation to DOT, you must do the following.
• Measure the freezing point of pure water.
• Measure the freezing-point depression for each solution.
• Calculate the molality of particles.

## Safety

Always wear safety goggles and a lab apron to protect your eyes and clothing. If you get a chemical in your eyes, immediately flush the chemical out at the eyewash station while calling to your teacher. Know the locations of the emergency lab shower and eyewash station and the procedure for using them.

Do not touch any chemicals. If you get a chemical on your skin or clothing, wash the chemical off at the sink while calling to your teacher. Make sure you carefully read the labels and follow the directions on all containers of chemicals that you use. Do not taste any chemicals or items used in the laboratory. Never return leftovers to their original containers; take only small amounts to avoid wasting supplies.

Always clean up the lab and all equipment after use, and dispose of substances according to proper disposal methods. Wash your hands thoroughly before you leave the lab after all lab work is finished.

## Materials

- De-icer F-38 (frozen solution)
- De-icer H-22 (frozen solution)
- De-icer J-27 (frozen solution)
- Ice (frozen distilled water)
- 250 mL beaker
- 25 mL graduated cylinder
- Glass stirring rod
- Test tubes (large), 3
- Test-tube rack
- Thermometer, non-mercury

**Probe option**
- Thermistor probe

## Preparation

1. **Organizing Data**
   Create a table in your lab notebook with two columns and five rows. Label the first and second boxes of the first row *Substance* and *Temp. (°C)*. Label the remaining boxes of the first column $H_2O$, *F-38*, *H-22*, and *J-27*.

## Technique

2. Fill a 250 mL beaker with crushed ice. Add distilled water to make a slurry of ice and water.

3. Measure the temperature of the ice-water mixture by holding a thermometer or a thermistor probe in the mixture. Be sure to keep the instrument off the bottom of the beaker. Wait until the temperature becomes constant, and then read and record the freezing temperature of pure water to the nearest 0.1°C in your data table.

4. Obtain a piece of frozen *F-38* solution that is about 25 mL in volume. Wrap it in toweling, and smash the toweling on the floor two or three times until it has the consistency of crushed ice. Pour the fragments into a test tube.

5. Allow the solid in the test tube to melt sufficiently so that liquid and solid are both present. Stir with a stirring rod to mix the solid and liquid. After stirring, hold a thermometer or a thermistor probe in the mixture but do not allow it to rest on the bottom of the test tube. When the temperature is constant, read and record the freezing point of the *F-38* solution.

6. Rinse the thermometer and the stirring rod thoroughly. Dry each with a paper towel.

7. Repeat steps **4–6** for the frozen *H-22* and *J-27* solutions. Record the freezing points of these two solutions in your data table.

## Cleanup and Disposal

8. Each solution should be returned to the disposal container that your teacher set out for it. Clean all apparatus and your lab station. Remember to wash your hands thoroughly before leaving the laboratory.

## Analysis and Interpretation

1. **Organizing Data**
   Determine the freezing-point depression, $\Delta t_f$, for each solution. The freezing-point depression is the difference between the freezing temperature of the solution and the freezing temperature of the pure solvent, water.

2. **Analyzing Data**
   From the freezing-point depression and the value of the molal freezing-point constant, calculate the "effective molality," which is the molality multiplied by the number of moles of particles per mole of solute ($m \times n$) of each solution, using the equation from the Background section.

3. **Analyzing Methods**
   Explain how you can be sure the temperatures you measured were the melting/freezing points of the solutions tested.

4. **Analyzing Methods**
   Would your results change if you used beaker-sized blocks of the frozen solutions? Why or why not?

## Conclusions

5. **Evaluating Conclusions**
   The Department of Transportation would like to use as little solute as possible on the roads. They want the greatest depression of freezing point per gram of solute. Which of the three solutes represents the best option for the DOT?

6. **Evaluating Conclusions**
   The Department of Transportation is also interested in the costs involved. Given the unit costs shown in the table below, which substance is the most economical? (Hint: consider which provides the greatest freezing-point depression per dollar.)

| Substance | Cost per kg |
|-----------|-------------|
| F-38 | $7.85 |
| H-22 | $6.20 |
| J-27 | $9.30 |

7. **Evaluating Methods**
   Why was it important to keep the thermometer off the bottom of the beaker in steps **3** and **5**?

## 8. Evaluating Methods

Why was it necessary to measure the temperature of the ice and pure water mixture, instead of assuming it to be 0.0°C?

## Extensions

### 1. Applying Conclusions

The solute in F-38 has the formula $MX_2$, and forms three particles on dissolving in water. The solute in H-22 has the formula MX, and forms two particles on dissolving. The solute in J-27 is a molecular solute which does not dissociate. Calculate the apparent molar mass for each solute. (Hint: first calculate the molal concentration of each solution using the information given here to determine $n$. Because each solution was made from 100.0 g of solute dissolved in 1.0 kg of solvent, you can then determine approximate molar masses.)

### 2. Applying Models

Will a 1.0 $m$ copper(II) sulfate solution have the same freezing-point depression as a 1.0 $m$ copper(II) chloride solution?

### 3. Connecting Ideas

Calculate the number of water molecules in enough 1.0 $m$ aqueous solution of sucrose to contain 2.0 mol of sucrose.

### 4. Designing Experiments

What possible sources of error can you identify with this procedure? If you can think of ways to eliminate them, ask your teacher to approve your plan, and run the procedure again.

### 5. Research and Communication

Not everyone believes that using salts as a de-icer on roads is a good idea. The subject is controversial in many northern towns where snowfall is heavy. Find out about the pros and cons. When you have considered all points of view, either hold a class debate, or write a personal position paper and present it as an oral report.

### 6. Applying Ideas

Ethylene glycol, $C_2H_6O_2$, is a molecular compound used as an antifreeze in automobile engines. What is the molal concentration of ethylene glycol in engine cooling water that will protect an engine when the temperature falls to $-30°C$?

### 7. Applying Ideas

Antifreeze containing ethylene glycol is also needed in automobile engines in extremely hot climates. Using what you know about other colligative properties, explain why this is so.

### 8. Relating Ideas

Different solvents have different freezing-point depression constants. For benzene, $K_f$ is $-5.12°C/m$. Pure benzene freezes at 5.50°C. What would be the freezing temperature of a 0.750 $m$ solution of sucrose in benzene?

# Freezing-Point Depression— Making Ice Cream

April 6, 1995

Ms. Martha Li-Hsien
Director of Research
CheMystery Labs, Inc.
52 Fulton Street
Springfield, VA 22150

Dear Ms. Li-Hsien:

In our business, it takes more than a great ice cream recipe to be successful. As you may know, when ice cream is made, it must be kept at a temperature below 0°C. Most ice-cream makers achieve these conditions using a saltwater and ice mixture packed around the container.

With our recent acquisition by D & J Food Processing, many of their products are available to us at a reasonable cost, including chemicals such as calcium acetate, $Ca(C_2H_3O_2)_2 \cdot 2H_2O$, and sodium hydrogen carbonate, $NaHCO_3$.

Please investigate the use of these chemicals and sodium chloride, $NaCl$, in saturated solutions. Determine the solubility of each compound, the molar and molal concentrations, the theoretical freezing-point depression for each compound, and the actual temperature of a mixture of the solution and ice.

Sincerely,

*Caitlin Cunningham*

Caitlin Cunningham, President
Bubba and Hugo's Ice Cream
A Division of D & J Food Processing

# Memorandum

*Date:* April 7, 1995
*To:* Vanessa Fanning
*From:* Martha Li-Hsien

CheMystery Labs, Inc.
52 Fulton Street
Springfield, VA 22150

D & J Food Processing could be a great prospect for future contracts. Our capabilities are a good match for their needs, so be sure to perform a careful and thorough analysis so that we will be assured of repeat business with them.

Mr. Cunningham has been negotiating with our lawyers. Instead of our usual arrangement, they are offering us $15 000 plus expenses for our analysis. Before we can sign the contract, we need the following items from you.

- detailed one-page summary of your procedure
- data charts and tables to organize your information
- itemized expense sheet including total expenditures

As soon as you get approval, begin your work. After your investigation, prepare a report in the form of a two-page letter to Mr. Cunningham at Bubba and Hugo's. Be certain that your report includes the following items.

- theoretically calculated and experimentally measured freezing-point depressions for 50.0 mL of saturated solution for each solute
- percentage difference between the calculated and measured freezing-point depressions
- molal and molar concentrations of the saturated solutions
- values of $K_{sp}$ for each compound
- detailed invoice for $15 000 more than actual total expenditures (be sure to include all expenses so that we will be reimbursed)

## Required Precautions

 Always wear safety goggles and a lab apron to protect your eyes and clothing.

 The ice mixture can cause frostbite if you handle it for too long. Use tongs, towels, or insulated gloves when handling the containers.

 Confine any long hair or loose clothing. Use tongs to handle any glassware or equipment that has been heated.

 Do not touch any chemicals. If you get a chemical on your skin or clothing, wash the chemical off at the sink while calling to your teacher.

 Always wash your hands thoroughly when finished.

## References

For this lab you will need to refer to your textbook for information about the following topics: saturated solutions, molarity, molality, colligative properties, equilibrium systems, solubility equilibria, and the equilibrium constant, $K_{sp}$.

Note that this Investigation requires skills that you used in measuring freezing-point depression and in measuring the solubility product constant for a dissolved substance.

In order to determine molarity and molality of the saturated solutions, it will be necessary to determine the number of moles of each compound, the volume of solution containing that amount, and the mass of the solvent that dissolves that amount.

Base your disposal costs on the mass used for each solute.

## Spill Procedures/ Waste Disposal Methods

- In case of spills, follow your teacher's instructions.
- Discard the solutions from this lab by flushing them down the drain.
- Clean the area and all equipment.

## MATERIALS for D&J/Bubba and Hugo's Ice Cream

### REQUIRED ITEMS
(You must include all of these in your budget.)

| | |
|---|---|
| Lab space/fume hood/utilities | 15 000/day |
| Standard disposal fee | 2 000/g |
| Balance | 5 000 |
| Beaker tongs | 1 000 |
| Crucible tongs | 2 000 |

### REAGENTS AND ADDITIONAL EQUIPMENT
(Include only what you'll need in your budget.)

| | |
|---|---|
| $Ca(C_2H_3O_2)_2 \cdot 2H_2O$ | 500/g |
| NaCl | 500/g |
| $NaHCO_3$ | 500/g |
| Ice | 500 |
| 250 mL beaker | 1 000 |
| 400 mL beaker | 2 000 |
| 250 mL flask | 1 000 |
| 25 mL graduated cylinder | 1 000 |
| 100 mL graduated cylinder | 1 000 |
| Bunsen burner and related equipment | 10 000 |
| Desiccator | 3 000 |
| Evaporating dish | 1 000 |
| Glass stirring rod | 1 000 |
| Hot plate | 8 000 |
| Litmus paper | 1 000/piece |
| Paper clips | 500/box |
| Plastic foam cup | 1 000 |
| Ring stand/ring/wire gauze | 2 000 |
| Stopwatch | 2 000 |
| Probe | 2 000 |
| Thermometer, nonmercury | 2 000 |
| Toothpicks | 500/dozen |
| Wash bottle | 500 |

*No refunds on returned chemicals or unused equipment.

### FINES

| | |
|---|---|
| OSHA safety violation | 2 000/incident |

# Measuring pH—Home Test Kit

April 20, 1995

Mr. Reginald Brown
Director of Materials Testing
CheMystery Labs, Inc.
52 Fulton Street
Springfield, VA 22150

Dear Mr. Brown:

Consumers are becoming more aware of the chemicals around their homes, and they want more information about them. They are asking for home chemical test kits to examine water quality in pools and fish tanks or to check the lead content of a painted wall. Our company would like to tap into this market by developing an inexpensive and easy-to-use pH test kit.

Because any test kit for the home must use very safe materials, we are interested in acid-base indicators that are made from natural products that consumers will recognize. Our kit must cover the pH range from 2 to 12 and be easy to use and interpret. We would like your company to research a wide range of natural indicators and recommend a product for us to test-market within the next six months.

In return, we will pay all reasonable expenses, an additional fee of $25 000, and an option for $\frac{1}{2}$% of the royalties on the first five years' sales of this product.

Sincerely,

*Victor Perkes*

Victor Perkes
Vice-President
Home Testing Products

# Memorandum

*Date:* April 21, 1995
*To:* Benito Ramirez
*From:* Reginald Brown

*CheMystery Labs, Inc.*
*52 Fulton Street*
*Springfield, VA 22150*

If we are successful in this project and the pH test kit sells well, we could be asked to help this start-up company develop other test kits. Unfortunately, our budget is a little tight, and we can afford an up-front investment of only $150 000.

Before you begin, I need the following items to present to the company's directors in order to authorize the $150 000 investment.

- detailed one-page plan for your procedure with all necessary data tables
- list of items you will test
- list of the equipment and materials you will need along with the total cost

To get started, pick at least six likely substances to test. I recommend using parts of plants, such as edible fruits and vegetables. No charge will be made for these materials, but you must gather them yourself. For each one, take enough solid starting material to fill a 400 mL beaker to the 250 mL line. Add enough water to fill the beaker. Heat the mixture at about 70°C for about 15 min. Allow it to cool before proceeding. Test the indicator properties of the liquid that remains.

Begin working when your plan is approved. When your team has the answers, prepare a report in the form of a two-page letter to Victor Perkes. The letter must include the following.

- description of the proposed product (solutions, strips, or another method), including your recommendation of the three indicators to be used in the product that together will provide the broadest range and most accurate values
- list of all indicators tested
- detailed and organized data table
- summary paragraph describing your rationale for choosing the indicators
- suggestions for scaling up the manufacturing of the test kit
- suggestions for the wording of instructions for the test kit
- detailed invoice for all expenses and services

## Required Precautions

 Always wear safety goggles and a lab apron to protect your eyes and clothing.

 Do not touch any chemicals. If you get a chemical on your skin or clothing, wash the chemical off at the sink while calling to your teacher.

 Acids and bases are corrosive. If any gets on your skin, wash the area immediately with running water while calling your teacher.

 Confine any long hair or loose clothing. Use tongs to handle any glassware or equipment that has been heated. Allow the heated solution to cool before decanting it or testing its indicator properties.

 Always wash your hands thoroughly when finished.

## Spill Procedures/ Waste Disposal Methods

- In case of spills, follow your teacher's instructions.
- Dispose of solid wastes in the container indicated by your teacher. The liquid wastes should be tested for pH. Acidic, neutral, and basic wastes should each go in their own containers.
- Clean the area and all equipment after use.
- Wash your hands when you are finished working in the lab.

## MATERIALS for Home Testing Products
### REQUIRED ITEMS
*( You must include all of these in your budget.)*

| | |
|---|---|
| Lab space/fume hood/utilities | 15 000/day |
| Standard disposal fee | 2 000/L of solution |
| Beaker tongs | 1 000 |
| Buret clamp, double | 1 000 |
| Buret tubes, 2 | 10 000 |

### REAGENTS AND ADDITIONAL EQUIPMENT
*( Include in your budget only what you'll need.)*

| | |
|---|---|
| 1.0 M NaOH | 100/mL |
| 1.0 M HCl | 100/mL |
| 250 mL beaker | 1 000 |
| 400 mL beaker | 2 000 |
| 250 mL Erlenmeyer flask | 1 000 |
| 100 mL graduated cylinder | 1 000 |
| Bunsen burner/related equipment | 10 000 |
| Glass stirring rod | 1 000 |
| Hot plate | 8 000 |
| pH paper | 2 000/piece |
| pH meter | 3 000 |
| pH probe | 5 000 |
| Ring stand/ring/wire gauze | 2 000 |
| Ruler | 500 |
| Stopwatch | 5 000 |
| Test tube (large) | 1 000 |
| Thermistor probe | 2 000 |
| Thermometer, nonmercury | 2 000 |
| Wash bottle | 500 |

\* No refunds on returned chemicals or unused equipment.

### FINES

| | |
|---|---|
| OSHA safety violation | 2 000/incident |

## References

Reread the sections of your textbook that discuss the operation and purpose of visual indicators. Review the pH scale and its usefulness in relating the strength of an acid or base solution. The dyes in many naturally colored food products can be used as indicators. Red cabbages, berries, and beets contain pH-sensitive chemicals. Do not limit yourself to this list, but try other similar substances.

# Measuring pH—Acid Precipitation Testing

April 24, 1995

Sandra Fernandez
Director of Development
CheMystery Labs, Inc.
52 Fulton Street
Springfield, VA 22150

Dear Ms. Fernandez:

A research team recently discovered that nearby Preston Lake has a pH of 3.0, presumably due to a combination of acid precipitation and point-source pollution. Not surprisingly, many fish were found dead or dying, and the plankton count was far too low. We know that some adult fish begin to die if the pH drops below 5.0, and no fish can reproduce when the pH is below 4.5.

The state legislature recently passed a measure allowing the State Environmental Resources Commission to perform a test of lime neutralization of acidic lakes in the state. We have chosen Preston Lake as the test case.

We are authorized to offer you $100 000 to help us evaluate the use of lime, CaO, to neutralize the lake. Use the sample of lake water we are sending to determine the mass of lime needed for the first treatment of the lake to raise its pH to 6.0 or above. We estimate the volume of Preston Lake to be $2.5 \times 10^9$ L.

Sincerely,

*Teresa Flores*

Teresa Flores
Assistant Director for Water Resources Management
State Environmental Resources Commission

# Memorandum

**Date:** April 25, 1995
**To:** Erika Strasser
**From:** Sandra Fernandez

*CheMystery Labs, Inc.*
*52 Fulton Street*
*Springfield, VA 22150*

This project is a <u>promising one</u>. If we do a good job, and if the state legislature decides to fund a full-fledged lake remediation project, we could be called upon to help the State Environmental Resources Commission plan treatments for other lakes.

Before you begin work, it is important that you prepare the following for the staff at the Environmental Resources Commission to review.

- <u>estimate of the amount of lime</u> you will need to neutralize the 500 mL of lake water, assuming that its pH is 3.0
- <u>detailed one-page plan</u> for your procedure with data tables
- <u>list of necessary equipment and materials</u> with their total cost

After your plan has been approved, begin the lab work. When you have the answers, prepare a report in the form of a <u>two-page letter</u> to Teresa Flores at the State Environmental Resources Commission. The letter must include the following items.

- <u>mass of lime</u> (in kg) that should be used <u>to neutralize</u> the $2.5 \times 10^9$ L lake
- <u>summary paragraph</u> of how you arrived at your conclusion
- detailed, organized <u>analysis of results</u> with calculations
- organized <u>table of data</u>
- <u>balanced chemical equations</u> for reactions of CaO with the acidic lake water
- discussion of any <u>possible sources of error</u> or <u>difficulties</u> in implementing the plan
- <u>itemized invoice</u> for equipment and services

## Required Precautions

 Always wear safety goggles and a lab apron to protect your eyes and clothing.

 Do not touch any chemicals. If you get a chemical on your skin or clothing, wash the chemical off at the sink while calling to your teacher.

 Acids and bases are corrosive. If any gets on your skin, wash the area immediately with running water while calling your teacher.

 Always wash your hands thoroughly when finished.

## Spill Procedures/ Waste Disposal Methods

- In case of spills, follow your teacher's instructions.
- Dispose of solid wastes in the container indicated by your teacher. The liquid wastes should be tested for pH. Acidic, neutral, and basic wastes should each go in their own separate containers.
- Clean the area and all equipment after use.
- Wash your hands when you are finished working in the lab.

## References

Acid precipitation often results from the combustion of fossil fuels, a process that can release oxides of nitrogen and sulfur into the atmosphere. These oxides are acid anhydrides. A discussion of the reactions of acid anhydrides with water can be found in the textbook. Review the pH scale and methods of measuring pH. Because of the small amount of sample being tested, it will be most effective if you measure the mass of a beaker with lime before you begin neutralizing. When the pH has reached an appropriate level, record the mass of the beaker with the leftover lime.

## MATERIALS for State Environmental Resources Commission

### REQUIRED ITEMS
*(You must include all of these in your budget.)*

| | |
|---|---|
| Lab space/fume hood/utilities | 15 000/day |
| Standard disposal fee | 20 000/L of solution |
| Balance | 5 000 |

### REAGENTS AND ADDITIONAL EQUIPMENT
*(Include in your budget only what you'll need.)*

| | |
|---|---|
| CaO, lime | 5 000/g |
| 150 mL beaker | 1 000 |
| 250 mL beaker | 1 000 |
| 400 mL beaker | 2 000 |
| 600 mL beaker | 3 000 |
| 250 mL Erlenmeyer flask | 1 000 |
| 100 mL graduated cylinder | 1 000 |
| Desiccator | 3 000 |
| Filter paper | 500/piece |
| Glass stirring rod | 1 000 |
| Magnetic stirrer | 5 000 |
| Paper clips | 500/box |
| pH meter | 3 000 |
| pH paper | 2 000/piece |
| pH probe | 5 000 |
| Rubber stopper | 1 000 |
| Ruler | 500 |
| Spatula | 500 |
| Tape | 500/10 cm |

*" No refunds on returned chemicals or unused equipment.*

### FINES

| | |
|---|---|
| OSHA safety violation | 2 000/incident |

HOLT
ChemFile
LAB PROGRAM

# Acid-Base Titration—Vinegar Tampering Investigation

April 28, 1995

Ms. Marissa Bellinghausen
Director of Investigations
CheMystery Labs, Inc.
52 Fulton Street
Springfield, VA 22150

Dear Ms. Bellinghausen:

Our company manufactures vinegar by the bacterial fermentation of alcohol. This process is capable of producing vinegar containing up to 12% acetic acid, but our practice is to dilute it to a concentration of about 5%, the usual concentration for food products.

Last week, the contract with one of the unions at the plant expired. For a short time, it appeared a strike was imminent, but it was averted at the last minute after an all-night negotiating session. The union has told us they believe that during the uncertainty one of the workers violated plant and union rules by tampering with the dilution equipment. The worker in question left without giving notice after the settlement was announced.

In the meantime, the union and the company agree that it is important to improve our quality assurance to protect consumers from any possible tampering. We would like you to evaluate the sample of vinegar from the production line and send us a report as soon as possible concerning its concentration.

Sincerely,

*David Gratton*

David Gratton
Director of Quality Control
Indian Summer Vinegar
A Division of D&J Foods

# Memorandum

Date: May 1, 1995
To: Michael Bennet
From: Marissa Bellinghausen

CheMystery Labs, Inc.
52 Fulton Street
Springfield, VA 22150

We are building a good relationship with D&J Foods. They're counting on us for quick, accurate results. This time, they are offering us $100 000.

I suggest that you titrate the vinegar. You can choose either a microscale or standard scale titration, but because microscale titrations involve smaller volumes, there is a greater chance for error. Before you start, give me the following items to review.

- detailed one-page plan for the procedure with all necessary data tables
- list of necessary equipment and materials with the total cost

Get started in the lab as soon as your plan has been approved. When you're finished, prepare a report in the form of a two-page letter to David Gratton. The letter must include the following.

- percentage by mass of acetic acid in the sample
- mass of acetic acid in 1.00 L of the sample
- average molarity of the vinegar, based on multiple trials
- pH of the vinegar sample
- $K_a$ of acetic acid
- comparison of and percent difference in your calculation and the accepted value for $K_a$ of acetic acid at 25°C, $1.76 \times 10^{-5}$
- summary paragraph describing how you arrived at your answers and discussing any difference between the experimental and accepted values for $K_a$
- data tables and all calculations
- balanced chemical equation for the titration reaction
- detailed invoice for all materials and services

## Required Precautions

 Always wear safety goggles and a lab apron to protect your eyes and clothing.

 Do not touch any chemicals. If you get a chemical on your skin or clothing, wash the chemical off at the sink while calling to your teacher.

 Acids and bases are corrosive. If any gets on your skin, wash the area immediately with running water while calling your teacher.

 Always wash your hands thoroughly when finished.

## Spill Procedures/ Waste Disposal Methods

- In case of spills, follow your teacher's instructions.
- The liquid wastes should be tested for pH. Acidic, neutral, and basic wastes should each go in their own separate containers.
- Clean the area and all equipment after use.
- Wash your hands when you are finished working in the lab.

## MATERIALS for Indian Summer Vinegar

### REQUIRED ITEMS
*(You must include all of these in your budget.)*

| | |
|---|---|
| Lab space/fume hood/utilities | 15 000/day |
| Standard disposal fee | 20 000/L of solution |

### REAGENTS AND ADDITIONAL EQUIPMENT
*(Include in your budget only what you'll need.)*

| | |
|---|---|
| 1.0 M HCl | 500/mL |
| 1.0 M NaOH | 500/mL |
| Phenolphthalein | 2 000 |
| 150 mL beaker | 1 000 |
| 250 mL beaker | 1 000 |
| 125 mL Erlenmeyer flask | 1 000 |
| 25 mL graduated cylinder | 1 000 |
| 100 mL graduated cylinder | 1 000 |
| Aluminum foil | 1 000/cm² |
| Buret tube | 5 000 |
| Filter paper | 500/piece |
| Glass stirring rod | 1 000 |
| pH meter | 3 000 |
| pH paper | 2 000/piece |
| pH probe | 5 000 |
| Pipet or medicine dropper | 1 000 |
| Ring stand with buret clamp | 2 000 |
| Stopwatch | 5 000 |
| Test tube (large) | 1 000 |
| Wash bottle | 500 |
| Watch glass | 1 000 |

\* No refunds on returned chemicals or unused equipment.

### FINES

| | |
|---|---|
| OSHA safety violation | 2 000/incident |

## References

Titration, a way of finding the concentration of an acid solution, is discussed in your textbook. Review this section for examples of how the molarity of an acid can be calculated from titration data, and for determining the acid dissociation constant, $K_a$, of an acid using the pH and molarity of the solution. To determine the mass percentage of acetic acid in vinegar, consider the density of the vinegar to be 1.01 g/mL.

# Catalysts

## Objectives

*Observe* the decomposition of hydrogen peroxide.

*Measure* the time needed for the production of a given volume of product.

*Graph* experimental data, and determine the reaction rate.

*Infer* the effect of a catalyst on the rate of the reaction.

*Determine* the effect of the concentration of hydrogen peroxide on the reaction rate.

## Situation

The winter production of your town's Drama Club will be a stage adaptation of Snow White. In one scene, a large glass flask, filled with what appears to be water, is supposed to begin bubbling and boiling instantly and without a source of heat when the wicked witch drops in a pinch of powdered "wing of bat." The director is looking for a simple, safe chemical reaction that produces a gas and can be initiated by dropping a pellet into a colorless liquid. She has asked you to investigate the decomposition of hydrogen peroxide, $H_2O_2$, catalyzed by sodium iodide, NaI. She needs data about the reaction rate and how the rate is related to the concentration of both hydrogen peroxide and sodium iodide so that she can adjust it for an effect that will be visible to the audience but safe for everyone on stage.

## Background

The measurement of reaction rates and the effects of a catalyst on the rate of reaction are discussed in the textbook. The decomposition of hydrogen peroxide to water and oxygen gas is a good illustration.

$$2H_2O_2(aq) \longrightarrow 2H_2O(l) + O_2(g)$$

Without a catalyst, the reaction proceeds slowly, but the rate of reaction is increased dramatically when a catalyst is used. By varying the concentrations of either the reactant or the catalyst, the dependence of the rate on these concentrations can be determined. Rate is measured by determining either the decrease in concentration of a reactant with time or the increase in concentration of a product with time. In this experiment you will use sodium iodide, NaI, as the catalyst and measure the time needed to generate a certain volume of the oxygen product when different concentrations of $I^-$ or $H_2O_2$ are used.

## Problem

To provide the Drama Club with the information they need, you must do the following.

• Carry out the reaction using different concentrations of $I^-$ and $H_2O_2$.

• Determine the times needed to produce equal volumes of oxygen for different concentrations.

• Graph the data to determine the rate.

• From the graphs, identify a rate law that indicates how the rate is affected by changes in the concentration of $H_2O_2$ and of the $I^-$ catalyst.

## Safety

Always wear safety goggles and a lab apron to protect your eyes and clothing. If you get a chemical in your eyes, immediately flush the chemical out at the eyewash station while calling to your teacher. Know the locations of the emergency lab shower and eyewash station and the procedure for using them.

Do not touch any chemicals. If you get a chemical on your skin or clothing, wash the chemical off at the sink while calling to your teacher. Carefully read the labels and follow the directions on all containers of chemicals that you use. Do not taste any chemicals or items used in the laboratory. Never return leftovers to their original containers; take only small amounts to avoid wasting supplies.

Never put broken glass in a regular waste container. Broken glass should be disposed of separately according to your teacher's instructions.

Always clean up the lab and all equipment after use, and dispose of substances according to proper disposal methods. Wash your hands thoroughly before you leave the lab after all lab work is finished.

## Preparation

**1. Organizing Data**

Prepare a data table with 4 columns and 17 rows. In the first row, label the boxes in the second through fourth columns *Rxn 1 (s)*, *Rxn 2 (s)*, and *Rxn 3 (s)*. In the second through fourth rows of the first column, label the boxes *NaI (g)*, *3% $H_2O_2$ (mL)*, and *$H_2O$ (mL)*. Fill in the remaining 13 rows of this column with even numbers of milliliters of $O_2$ from 2 to 26. The three remaining columns provide space for recording time in seconds for three reactions. You will also need a space to record the initial volume of the buret for each of the three reactions and several spaces for recording the initial temperature of the water in the pneumatic trough or plastic tub.

**2.** On separate pieces of weighing paper, measure the mass of three samples of NaI to the nearest 0.01 g. The sample for reaction 1 should be about 0.10 g. The sample for reaction 2 should be about 0.20 g. The sample for reaction 3 should be about 0.10 g. Label each sample with the number of the reaction in which it will be used, and record the masses in your data table.

## Technique

**3.** Using a buret clamp, attach the buret to the ring stand. Attach one piece of rubber tubing to the base of the buret. Thread the rubber tubing through a ring that is attached to the ring stand, and then attach the leveling bulb to the other end of the rubber tubing, as shown in Figure A on the next page.

## Materials

- 3% $H_2O_2$ solution, 125 mL
- NaI, 0.50 g
- 125 mL Erlenmeyer flask
- 100 mL graduated cylinder
- 100 mL Mohr buret
- Balance
- Buret clamp
- Leveling bulb
- Medicine droppers, 2
- Pneumatic trough or plastic tub
- Ring stand and ring
- Rubber tubing, 2 pieces
- Rubber stopper, one-hole (for Erlenmeyer flask)
- Rubber stopper, one-hole (for Mohr buret)
- Stopwatch or clock with second hand
- Thermometer, non-mercury
- Weighing paper, 3 pieces

**Probe Option**
- Thermistor probe

Leveling bulb (Note: keep level of fluid inside bulb the same as the level in buret when reading volumes of gas)

Medicine dropper inserted in stopper

Water level in Mohr buret

Rubber tubing

Rubber tubing

Medicine dropper inserted in stopper

125 mL Erlenmeyer flask with reaction mixture

Pneumatic trough or tub

**Figure A**

4. Fill the pneumatic trough or tub with water at room temperature. Add room-temperature water to the buret until the water is about 10 mL from the top when the water in the leveling bulb is at the same level as the water in the buret. To maintain constant pressure throughout the experiment, the water in the leveling bulb should be kept at the same level as the water in the buret. (The bulb may rest in the ring clamp while you complete the assembly of the apparatus.)

5. Your teacher should have already inserted the medicine droppers into the one-hole stoppers. Attach a second piece of rubber tubing, with one of the medicine droppers and stoppers on either end. Insert one stopper into the top of the buret and the other into the top of the flask, as shown in Figure A.

6. Check the system for leaks by lowering and raising the leveling bulb. There should be only small changes as the leveling bulb is lowered. If there are large changes, the system has a leak that must be corrected.

7. Record the temperature of the water in the pneumatic trough or plastic tub.

8. Using a graduated cylinder, measure about 50 mL of 3% $H_2O_2$. Measure the actual volume to the nearest 0.1 mL, and record it in your data table.

9. Record the initial volume of the buret, and decide how you and your partner will divide the tasks in step **10**.

10. Remove the stopper assembly from the flask, and place the NaI for reaction 1 into it. Add the hydrogen peroxide, and quickly replace the stopper assembly. One member of your team should swirl the flask throughout the reaction and watch the time. The other member should keep the water in the leveling bulb at the same level as the water in the buret and watch the volume. Begin recording the time and volume after approximately 6 mL of oxygen gas have evolved.

11. Take time and volume readings after the evolution of each 2 mL of gas until you have about 13 readings.

12. Thoroughly rinse the Erlenmeyer flask. Check the temperature of the water in the trough, and, if necessary, adjust it to the temperature recorded in step **7** by adding small amounts of cold or warm water.

13. Repeat steps **8–12** for reaction 2, using 50 mL of 3% $H_2O_2$ and 0.20 g NaI.

14. Repeat steps **9–11** for reaction 3, but this time add 25 mL of distilled water to the 0.10 g NaI in the flask, and swirl it in the trough until the NaI is completely dissolved and the solution has had time to reach the same temperature as the water bath. Then add 25 mL of 3% $H_2O_2$, and proceed as before.

## Cleanup and Disposal

15. All solutions may be washed down the drain once they have stopped bubbling. Any excess solid NaI should be placed in the disposal container designated by your teacher. Clean up all equipment and your lab station. Remember to wash your hands.

## Analysis and Interpretation

1. **Evaluating Data**
   List the five factors that influence the rate of a reaction, and identify those that were studied in this experiment.

2. **Organizing Data**
   Plot your results for reactions 1, 2, and 3 on one graph, with the volume of $O_2$ in milliliters on the y-axis and the time in seconds on the x-axis. Draw the line of best fit through the points for each reaction.

3. **Interpreting Graphs**
   Calculate the slope for each line. How does it relate to the reaction rate? (Hint: be sure to include the appropriate units.)

4. **Evaluating Methods**
   Why was it important to check the water temperature and keep the bulb's water level the same as the water level in the Mohr buret? (Hint: recall what you know about pressure from the textbook.)

## Conclusions

5. **Inferring Conclusions**
   Use your graph and your calculated slopes for reactions 1 and 2 to determine how doubling the concentration of the catalyst, NaI, affected the rate of the reaction.

6. **Inferring Conclusions**
   Consider reactions 1 and 3. What effect did the change in concentration of hydrogen peroxide have on the rate of the reaction? (Hint: explain the effect quantitatively.)

7. **Applying Conclusions**
   Write a rate-law expression for this equation using the concentrations of $H_2O_2$ and $I^-$. (Hint: use the best value to within 0.1 for the exponents of the concentrations.)

**Chem**File

**8. Evaluating Data**

In this experiment you are studying the rate of a reaction by measuring the increase in volume of oxygen per second rather than the increase in concentration. Explain why volumes of gas, measured under the same conditions of temperature and pressure, can be used instead of concentrations in a rate equation. (Hint: consider what you know about gases from the textbook.)

**9. Applying Models**

Using collision theory, explain why increasing the concentration of a reactant often increases the rate of the reaction.

**10. Applying Models**

Under what circumstances might increasing the concentration of a reactant have no effect on the rate of a reaction?

**11. Applying Conclusions**

What concentrations of reactants do you recommend that the Drama Club use for the scene, given that they will be using a 1.0 L round-bottom flask?

## Extensions

**1. Designing Experiments**

What difficulties or sources of error can you identify with this procedure? If you can think of ways to eliminate them, ask your teacher to approve your plan, and run the procedure again.

**2. Analyzing Data**

The following data were collected in another hydrogen peroxide experiment. What is the average rate of reaction for each of the periods between measurements?

| Volume $O_2$ (mL) | 0.0 | 2.0 | 3.8 | 5.9 | 7.9 |
|---|---|---|---|---|---|
| Time (s) | 0.0 | 60.0 | 115.0 | 180.0 | 245.0 |

**3. Predicting Outcomes**

Predict what the rate might be if the reaction in item **2** continued until 10 mL of $O_2$ were formed. Explain your prediction.

**4. Connecting Ideas**

Explain in terms of activation energy for a reaction how a catalyst operates to increase the rate of a reaction.

**5. Applying Ideas**

Consider the five factors that control the rate of a reaction, and explain why there is a danger of explosion in saw mills and grain elevators, where large amounts of dry, powdered, combustible materials might be present in the air.

**6. Research and Communication**

Some metabolic functions produce hydrogen peroxide as a byproduct. When present in high concentration, it can cause severe damage to tissue. Find out the mechanisms your body uses to deal with this potentially dangerous substance, and write a short paper summarizing your research.

# Catalysts—Peroxide Disposal

May 4, 1995

Mr. Reginald Brown
Director of Materials Testing
CheMystery Labs, Inc.
52 Fulton Street
Springfield, VA 22150

Dear Mr. Brown:

In the synthesis of one of our products, hydrogen peroxide is generated as a byproduct. Disposing of the $H_2O_2$ in the sewer system is not permitted because it can be harmful to wildlife and can cause excess pressure to build up in clogged pipes. We are considering decomposing the hydrogen peroxide into oxygen and water but need a catalyst to speed up the reaction.

We need your firm to investigate three likely catalysts that could be used to speed decomposition of $H_2O_2$: sodium iodide, manganese dioxide, and yeast. Sodium iodide and manganese dioxide are inorganic catalysts. Yeast is a fungus that produces enzymes which catalyze the decomposition.

Please identify the most effective and economical method for us to use. We are prepared to pay up to $100 000 for satisfactory work.

Sincerely,

*Christopher Sasseen*

Christopher Sasseen
Division Manager
Industrial Engineering Systems, Inc.

# Memorandum

**Date:** May 5, 1995
**To:** Theodore Cole
**From:** Reginald Brown

CheMystery Labs, Inc.
52 Fulton Street
Springfield, VA 22150

We've got to give this contract our best shot because Industrial Engineering Systems is a <u>huge conglomerate</u> and a series of contracts with them could give us the <u>stability we need</u> to further expand our customer base.

Before you make a plan, <u>check the solubility</u> of NaI and $MnO_2$ in a chemical handbook to be sure they will be satisfactory. Test a small amount of yeast to make sure it is satisfactory. Also be sure to <u>note the costs</u> of all three catalysts. Then, before you begin work in the lab, I will need to review the following items.

- <u>one-page plan</u> for your procedure, with <u>data tables</u>
- <u>detailed list</u> of the equipment and materials you need, with total cost

Begin the lab work as soon as your plan is approved. When you have the answers, prepare a <u>two-page letter</u> to Christopher Sasseen, Division Manager, Industrial Engineering Systems, Inc. Be sure to include the following information.

- <u>recommendation of which catalyst</u> is the most effective and why
- summary paragraph explaining <u>how your procedure solved this problem</u>
- <u>rates of decomposition</u> of $H_2O_2$ using the three catalysts
- <u>graphs</u> of your data
- detailed, organized <u>data table</u>
- <u>detailed invoice</u> for costs of materials and services

## Required Precautions

 Always wear safety goggles and a lab apron to protect your eyes and clothing.

 Do not touch any chemicals. If you get a chemical on your skin or clothing, wash the chemical off at the sink while calling to your teacher.

 Never put broken glass in a regular waste container. Broken glass should be disposed of separately. Put the glass pieces in the container designated by your teacher.

 Always wash your hands thoroughly when finished.

## Spill Procedures/ Waste Disposal Methods

- In case of spills, follow your teacher's instructions.
- $MnO_2$ and any $H_2O_2$-$MnO_2$ mixtures should be placed in a designated and labeled disposal container. **Do not try to wash $MnO_2$ down the drain.**
- Pure $H_2O_2$, yeast, and NaI solutions may be poured down the drain.
- Dispose of any leftover solids in the separate disposal containers designated by your teacher.
- Clean the area and all equipment after use.

## References

Refer to the textbook for a discussion of reaction rate, the dependence of the rate on the concentration of the reactants, and the effects of catalysts. Information on measuring the rate of a reaction and graphing rate data to determine the rate law can also be found in your textbook. The procedure used to measure the rate of decomposition of $H_2O_2$ will be very similar to one in an Exploration that you and your team recently completed.

## MATERIALS for Industrial Engineering Systems, Inc.

### REQUIRED ITEMS

*(You must include all of these in your budget.)*

| | |
|---|---|
| Lab space/fume hood/utilities | 15 000/day |
| Standard disposal fee | 20 000/L of solution |
| 100 mL Mohr buret | 5 000 |
| Balance | 5 000 |
| Buret clamp | 1 000 |
| Leveling bulb | 500 |
| Pneumatic trough or plastic tub | 1 000 |
| Ring stand/ring | 2 000 |

### REAGENTS AND ADDITIONAL EQUIPMENT

*(Include in your budget only what you'll need.)*

| | |
|---|---|
| NaI | 750/g |
| $MnO_2$ | 150/g |
| Yeast | 100/g |
| 3% $H_2O_2$ solution | 500/mL |
| Ice | 500 |
| 250 mL beaker | 1 000 |
| 100 mL graduated cylinder | 1 000 |
| 125 mL Erlenmeyer flask | 1 000 |
| Beaker tongs | 1 000 |
| Filter paper | 500/piece |
| Glass stirring rod | 1 000 |
| Rubber stopper with medicine dropper inserted | 1 000 |
| Rubber tubing | 500/cm |
| Stopwatch | 5 000 |
| Thermometer | 2 000 |
| Thermistor probe | 2 000 |
| Wash bottle | 500 |
| Weighing paper | 500/piece |

\* No refunds on returned chemicals or unused equipment.

### FINES

| | |
|---|---|
| OSHA safety violation | 2 000/incident |

**Chem**File

# Electroplating for Corrosion Protection

## Objectives

***Construct*** an electrolytic cell with an electrolyte, two electrodes, and a battery.

***Use*** the electrolytic cell to plate one metal onto another.

***Test*** the plated metals and a sample of the original metal to determine how well they resist corrosion by an acidic solution.

***Relate*** the results to the activity series and to standard reduction potentials.

## Situation

Your company has been contacted by a manufacturer of electrical circuits. The company uses 1.0 M HCl to clean newly manufactured circuits. They've decided to store the acid in large metal tanks instead of reagent bottles. The company has narrowed down the choices to copper, zinc-plated copper, and iron-plated copper. You have been asked to evaluate these choices.

## Background

Solutions of acids can oxidize some metals. The single displacement reaction of magnesium and hydrochloric acid is an example.

$$\underset{0}{Mg(s)} + \underset{+1 \ -1}{2HCl(aq)} \longrightarrow \underset{+2 \ -1}{MgCl_2(aq)} + \underset{0}{H_2(g)}$$

Metals that react can be electroplated with a thin layer of a less reactive metal. In an electroplating cell, electrical energy is used to reduce metal ions in solution, causing it to adhere to the surface of an object functioning as an anode. Figure A shown here summarizes the apparatus and the process.

Battery

$e^-$

Cathode: object to be plated (Fe)
$Ag^+ + e^- \longrightarrow Ag$
(added to Fe object)

Anode: plating metal (Ag)
$Ag \longrightarrow Ag^+ + e^-$ (removed from plating metal strip)

$AgNO_3$ solution

**Figure A**

## Problem

You must first prepare samples of the plated metals. Then you must test each one in 1.0 M HCl to determine which resists corrosion the best.

## Safety

Always wear safety goggles and a lab apron to protect your eyes and clothing. If you get a chemical in your eyes, immediately flush the chemical out at the eyewash station while calling to your teacher. Know the locations of the emergency lab shower and eyewash station and the procedure for using them.

Do not touch any chemicals. If you get a chemical on your skin or clothing, wash the chemical off at the sink while calling to your teacher. Make sure you carefully read the labels and follow the directions on all containers of chemicals that you use. Never return leftovers to their original containers; take only small amounts to avoid wasting supplies.

Call your teacher in the event of an HCl spill. Acid or base spills should be cleaned up promptly, according to your teacher's instructions.

When you use a Bunsen burner, confine any long hair and loose clothing. Do not heat glassware that is broken, chipped, or cracked. Use tongs or a hot mitt to handle heated glassware or other equipment because hot glassware does not look hot. If your clothing catches on fire, WALK to the emergency lab shower and use it to put out the fire.

Always clean up the lab and all equipment after use, and dispose of substances according to proper disposal methods. Wash your hands thoroughly before you leave the lab after all lab work is finished.

## Materials

- 1.0 M HCl, 50 mL
- $FeCl_3$ plating solution, 80 mL
- $ZnSO_4$ plating solution, 80 mL
- Copper wire, 10 cm lengths, 3
- Distilled water
- Iron strip, 1 cm × 8 cm
- Zinc strip, 1 cm × 8 cm
- 150 mL beakers, 3
- 400 mL beaker
- Balance
- Battery, 6 V (lantern type)
- Steel wool
- Stick-on label
- Stopwatch or clock with second hand
- Test-tube rack
- Test tubes, large, 3
- Wax pencil
- Wire with alligator clips, 2 pieces

### Optional equipment
- Beaker tongs
- Drying oven

## Preparation

**1. Organizing Data**

Prepare a data table in your lab notebook with four columns and six rows. Label the boxes in the second through fourth columns of the first row *Cu, Fe/Cu,* and *Zn/Cu.* Label the boxes in the second through sixth rows in the first column *Mass of wire (g), Plating time (s), New mass of wire (g), HCl time (s),* and *Mass of wire after HCl (g).* You will also need room to record your observations.

**2.** Label two of the 150 mL beakers *FeCl₃* and *ZnSO₄.* Label the 400 mL beaker *Waste.*

**3.** Make loops on one end of each of the wires. Using a piece of a stick-on label, label the wires *Cu, Fe/Cu,* and *Zn/Cu* just below the loops.

## Technique

**4.** Polish each wire and metal strip with steel wool.

**5.** Using the laboratory balance, measure the mass of the *Fe/Cu* wire. Record it in your data table.

**Chem**File

## Plating

6. Attach one end of a wire with alligator clips to the loop on the *Fe/Cu* wire. Attach the alligator clip on the other end of the wire to the negative (−) terminal of the battery.

7. Attach one end of the other wire with alligator clips to the iron strip. Clip the other end of this wire to the positive (+) terminal of the battery, as shown in Figure B below. Keep the iron strip away from the copper wire.

Alligator clip

Alligator clip

Battery

Alligator clip

Alligator clip

Iron strip

Fe/Cu copper wire

**Figure B**

8. Pour 80 mL of the $FeCl_3$ solution into the *$FeCl_3$* beaker.

9. Using a stopwatch or clock with second hand to measure the time, immerse both the copper wire and the iron strip in the beaker, being careful to keep the alligator clips out of the solution. Use the loop to hang the wire on the beaker.

10. After about 5 min, remove the copper wire and the iron strip. Record the time elapsed to the nearest 1.0 s.

11. Rinse the *Fe/Cu* wire with distilled water, collecting the rinse water in the *Waste* beaker. Record your observations about the *Fe/Cu* wire in your data table.

12. Hold the unplated *Zn/Cu* wire close to the plated *Fe/Cu* wire. Use a wax pencil to make a mark on the *Zn/Cu* and *Cu* wires at about the same level as the edge of the plating on the *Fe/Cu* wire. Measure and record the masses of these wires.

13. Repeat steps **6–11** with the zinc metal strip, zinc sulfate solution in the *$ZnSO_4$* beaker, and the wire labeled *Zn/Cu*. Be certain to plate the wire for exactly the same amount of time. Also be certain that the *Zn/Cu* wire is immersed up to the wax pencil mark.

14. Place the plated wires in the unlabeled 150 mL beaker so that they are not touching each other. Either allow the wires to dry in the beaker overnight, or place the beaker in a drying oven for 10 min. **Remember to use beaker tongs to handle all glassware that has been in the drying oven.**

15. After the wires have cooled, measure and record the masses of the *Fe/Cu* wire and the *Zn/Cu* wire in the data table.

16. In your data table rewrite the original mass of the *Cu* wire in the *New mass of wire (g)* row under the *Cu* column.

## Testing Reactivity

17. Fill each of the test tubes about one-third full of 1.0 M HCl. Place the test tubes in a test-tube rack.

18. Place one of the wires into each of the test tubes so that only the plated parts are in the HCl solution. Wait about 5 min.

19. Remove the wires, and rinse them with distilled water, collecting the rinse in the *Waste* beaker. Record the time and your observations about the wires in your data table. Place the wires in the unlabeled 150 mL beaker so that they do not touch. Dry the wires overnight, or place them in a drying oven for 10 min.

20. Remove the beaker of wires from the drying oven and allow it to cool. **Remember to use beaker tongs to handle beakers that have been in the drying oven.**

21. Measure and record the masses of the wires in your data table as *Mass of wire after HCl (g)*.

## Cleanup and Disposal

22. Your teacher will provide separate disposal containers for each solution, each metal, and contents of the *Waste* beaker. The HCl from the test tubes can be poured into the same container as the contents of the *Waste* beaker.

## Analysis and Interpretation

1. **Applying Models**
   Write the equation for the half-reactions occurring on the metal strip and the copper wire in the $FeCl_3$ and $ZnSO_4$ beakers. Which is the anode in each one? Which is the cathode in each one?

2. **Analyzing Data**
   How many grams and moles of iron and zinc were plated onto the *Fe/Cu* and *Zn/Cu* wires?

## Conclusions

3. **Analyzing Results**
   Which metal or metal combination was the least reactive in HCl? Explain the basis for your conclusion.

4. **Evaluating Conclusions**
   What disadvantages relating to the use of the least reactive metal for the tanks can you think of?

## Extensions

1. **Research and Communication**
   Find out what measures are taken to try to prevent metal bridges and buildings from corroding, and prepare a chart to show the different methods, their relative costs, and their general uses.

# Voltaic Cells

## Objectives

***Design*** **and construct various voltaic cells.**

***Measure*** **the actual voltages of the voltaic cells.**

***Calculate*** **the voltages of the voltaic cells using a table of standard reduction potentials.**

***Evaluate*** **cells by comparing actual cell voltages and voltages calculated from standard reduction potentials.**

***Write*** **equations for cell reactions by combining half-cell reactions.**

## Situation

NASA is requesting that interested consultants present proposals for providing electrical engineering services on their "Self-Propelled Robotic Planetary Explorer." The space agency asks that your company submit specifications for energy-producing cells that can be made easily from readily obtainable, inexpensive materials. Your proposal should provide information about the construction of the cells, the energy available from the cells, and the equations for the reactions occurring in the cells.

## Background

Redox reactions involve the transfer of electrons from a substance being oxidized to a substance being reduced. If the reactants are in contact, the energy released during the electron transfer will be in the form of heat. However, if the reactants are in separate half-cells connected by an external wire, the energy released is electrical energy in the form of electrons traveling through the wire. In electrochemical or voltaic cells, the oxidation and reduction half-reactions take place in separate half-cells, which often consist of a metal electrode immersed in a solution of its metal ions. The electrical potential, or voltage, that develops between the electrodes is a measure of the combined reducing strength of one reactant and oxidizing strength of the other reactant. This potential difference is measured in units of volts (V).

## Problem

NASA has provided guidelines for constructing and measuring the potential of a cell using copper and zinc electrodes. To prepare your proposal you will need to do the following.
• Construct the Cu-Zn voltaic cell, and measure the potential.
• Use standard reduction potentials to determine what other cells might be feasible using aluminum, zinc, and copper.
• Prepare two additional cells, and measure their potentials.
• Evaluate the possible choices by comparing the voltages of your cells with standard cell potentials.

## Safety

Always wear safety goggles and a lab apron to protect your eyes and clothing. If you get a chemical in your eyes, immediately flush the chemical out at the eyewash station while calling to your teacher. Know the locations of the emergency lab shower and eyewash station and the procedure for using them.

Do not touch any chemicals. If you get a chemical on your skin or clothing, wash the chemical off at the sink while calling to your teacher. Make sure you carefully read the labels and follow the directions on all containers of chemicals that you use. Do not taste any chemicals or items used in the laboratory. Never return leftovers to their original containers; take only small amounts to avoid wasting supplies.

Always clean up the lab and all equipment after use, and dispose of substances according to proper disposal methods. Wash your hands thoroughly before you leave the lab after all lab work is finished.

## Preparation

**1. Organizing Data**
Prepare a table for recording your data for three voltaic cells. In each of three columns, you will need a large row to draw a diagram of each cell and regular-sized rows to indicate the molarity of the solutions used and the measured voltage of the cells.

**2.** Remove any oxide coating from strips of aluminum, copper, and zinc by rubbing them with emery cloth. Keep the metal strips dry until you are ready to use them.

**3.** Label three 150 mL beakers $Al_2(SO_4)_3$, $CuSO_4$, and $ZnSO_4$. If you are using porous cups, also label a 250 mL beaker $ZnSO_4$ *half-cell*.

## Technique

**4.** If you are using a porous cup, pour 75 mL of 0.5 M $ZnSO_4$ into the $ZnSO_4$ *half-cell* beaker and 75 mL of 0.5 M $CuSO_4$ into one of the porous cups. If you are using a salt bridge, pour 75 mL of 0.5 M $ZnSO_4$ into the $ZnSO_4$ beaker and 75 mL of 0.5 M $CuSO_4$ into the $CuSO_4$ beaker.

**5.** If you are using a porous cup, place the cup containing $CuSO_4$ solution into the beaker containing the $ZnSO_4$ solution. If you are using a salt bridge, place one end into the solution in the $CuSO_4$ beaker and the other end into the solution in the $ZnSO_4$ beaker.

**6.** Place a zinc strip into the zinc solution and a copper strip into the copper solution.

## Materials

- 0.5 M $Al_2(SO_4)_3$, 75 mL
- 0.5 M $CuSO_4$, 75 mL
- 0.5 M $ZnSO_4$, 75 mL
- Aluminum strip, 1 cm × 8 cm
- Copper strip, 1 cm × 8 cm
- Zinc strip, 1 cm × 8 cm
- Distilled water
- 100 mL graduated cylinder
- Balance
- Emery cloth

**Porous Cup Option**
- 150 mL beakers, 3
- 250 mL beakers, 3
- Porous cup

**Salt Bridge Option**
- 150 mL beakers, 3
- Salt bridge

**Voltmeter Option**
- Voltmeter
- Wires with alligator clips, 2

**Probe Option**
- Voltage probe

7. Using the alligator clips, connect one wire to the end of the zinc strip and the second wire to the copper strip. Take the free end of the wire attached to the zinc strip, and connect it to one terminal on the voltmeter. Take the free end of the wire attached to the copper strip, and connect it to the other terminal on the voltmeter. The needle on the voltmeter should move to the right. If your voltmeter's needle points to the left, disconnect the wires and reconnect the zinc strip and copper strip to different terminals of the voltmeter. Immediately record the voltage reading in your data table, and disconnect the circuit.

8. Fill in the rest of the data table, including the sketch of the equipment as it appeared when connected.

9. If you are using porous cups, pour each of the solutions into the appropriately labeled 150 mL beakers.

10. Rinse the copper strip with a *very small* amount of distilled water, collecting the rinse in the $CuSO_4$ beaker. Rinse the zinc strip with a *very small* amount of distilled water, collecting the rinse in the $ZnSO_4$ beaker. If a porous cup was used, rinse the outside, collecting the rinse in the $ZnSO_4$ beaker. Then, rinse the inside of the porous cup, collecting the rinse in the $CuSO_4$ beaker. If a salt bridge was used, rinse each end into the corresponding beaker.

11. Use the table of standard reduction potentials in the textbook to calculate the standard voltages for the other cells you can build using copper, zinc, or aluminum. Build these cells and measure their potentials according to steps **4–10**.

## Cleanup and Disposal

12. Clean all apparatus and your lab station. Remember to wash your hands. Place the pieces of metal in the disposal containers designated by your teacher. Each solution should be poured in its own separate disposal container. Do not mix the contents of the beakers.

# Analysis and Interpretation

1. **Organizing Ideas**
   For each cell that you constructed, write the equations for the two half-cell reactions. From a table of standard reduction potentials, obtain the standard half-cell potentials for the half-reactions, and write these $E^0$ values after the equations.

2. **Organizing Ideas**
   Combine the two half-reactions for each cell you tested to obtain the equations for the total cell reactions.

3. **Organizing Ideas**
   Combine the $E^0$ values for each cell to find the standard potential for each cell.

4. **Resolving Discrepancies**
   Compare the actual cell voltages you measured with the standard cell voltages in item **3.** Explain why you would expect a difference. (Hint: consider how the construction of the standard cell differs from yours and the conditions under which standard potentials are measured.)

# Conclusions

5. **Inferring Conclusions**
   On the basis of the actual voltages that you measured, which cell would you recommend to NASA as the best energy producer?

6. **Applying Ideas**
   On the basis of your data, which metal is the strongest reducing agent? Which metal ion is the strongest oxidizing agent?

7. **Analyzing Methods**
   Why was it important that you removed any oxide coating from the metal electrodes before they were used?

8. **Applying Ideas**
   For each of the diagrams in your data table, indicate the direction of electron flow.

# Extensions

1. **Predicting Outcomes**
   Describe how and why the reactions would stop if the cells had been left connected.

2. **Evaluating Conclusions**
   As with any space flight, considerations of mass are critical for the missions in which NASA would use the "Self-Propelled Robotic Planetary Explorer." Given your answer to Extension **1,** which of the cells you tested would continue providing some electrical energy for the longest period, assuming that equal masses of the metals are used in all of the cells?

3. **Designing Experiments**
   Design a method that could use several of the electrochemical cells you constructed to generate more voltage than any individual cell provided. (Hint: consider what would happen if you linked an Al-Zn cell and a Zn-Cu cell. If your teacher approves your plan, test your idea.

### 4. Applying Ideas

The electrochemical cell shown below is at a temperature of 25°C. Identify the metal that will be oxidized if the current is allowed to flow.

Voltmeter

Wire

Wire

Salt bridge

Manganese
metal

Gold metal

$Au^{3+}$ solution

$Mn^{2+}$ solution

### 5. Research and Communication

Electric automobiles powered by rechargeable batteries are now in limited use in some areas. Find out about the available types of batteries, the range of the vehicle, and the advantages and disadvantages. Prepare a report to your class about your findings, including your predictions about the future of this form of transportation.

HOLT
# ChemFile
LAB PROGRAM

# Voltaic Cells—Designing Batteries

May 9, 1995

Ms. Sandra Fernandez
Director of Development
CheMystery Labs, Inc.
52 Fulton Street
Springfield, VA 22150

Dear Ms. Fernandez:

Rescue Adjuncts, Inc. makes a variety of survival and first-aid gear for outdoor enthusiasts. We want to develop a powerful signal light that can be used by people venturing into areas where ordinary modes of communication are not available. Life or death outcomes depend on whether a distress signal reaches observers who can send help.

We are requesting your help in developing voltaic cells capable of powering a signal light. Please research which of the following metals and their solutions will provide at least 5.0 V: aluminum, copper, magnesium, tin, and zinc.

Submit a proposal that includes the best prospects for the cells. Include a bid that estimates the cost of whatever research you need to do in order to recommend the voltaic cell that we hope will become indispensable to every adventurer.

Sincerely,

*Garlon Prewitt*

Garlon Prewitt
Chief Engineer
Rescue Adjuncts, Inc.

# Memorandum

*Date:* May 10, 1995
*To:* Lucinda Crevois
*From:* Sandra Fernandez

CheMystery Labs, Inc.
52 Fulton Street
Springfield, VA 22150

This job is by no means certain. My contacts elsewhere tell me that this offer went out to at least three of our competitors. If we want to get this business, we must be certain to have the best (and most cost-effective) proposal. As discussed in the letter from Garlon Prewitt, be sure your proposal includes the following.

- detailed one-page plan for your procedure
- combinations that are most likely to yield the highest voltages (maybe several cells linked together)

Voltmeter

Mg   Mg²⁺   Salt bridge   Cu²⁺   Cu   Mg   Mg²⁺   Salt bridge   Cu²⁺   Cu

- balanced redox equations and cell diagrams for the voltaic cells you will construct
- standard cell potentials for your proposed cells
- detailed data tables
- list of all necessary materials and equipment
- detailed bid with total costs (Because this is a bid, you need to factor in equipment, materials, labor, overhead, and profit.)

Begin working as soon as your bid is accepted. Make sure you keep your actual costs under the amount of the bid. When your work is finished, prepare a two-page report to Garlon Prewitt. Be sure to include the following items.

- your final recommendation for the cell or combination of cells that will provide 5.0 V
- sketch or diagram of the cell or combination
- paragraph summarizing your procedure
- detailed and organized data and analysis sections
- calculations showing what percentage difference there was between your measured cell potentials and the standard potentials calculated for the proposal
- invoice for the bid amount

## Required Precautions

Always wear safety goggles and a lab apron to protect your eyes and clothing. If you chose to work with the SnCl₂ solution, you must wear gloves because the solution irritates the skin.

Do not touch any chemicals. If you get a chemical on your skin or clothing, wash the chemical off at the sink while calling to your teacher.

Never put broken glass in a regular waste container. Broken glass should be disposed of separately. Put the glass pieces in the container designated by your teacher.

Always wash your hands thoroughly when finished.

## Spill Procedures/ Waste Disposal Methods

- In case of spills, follow your teacher's instructions.
- After testing a cell, rinse the metal strips (and the porous cup, if one was used) with a small amount of distilled water, collecting the rinse in the beaker containing the solution of that metal's ions.
- **Do not mix the solutions or pour them down the drain.** All solid and liquid wastes must be disposed of in their own separate containers, as designated by your teacher.
- Clean the area and all equipment after use.

## References

Review spontaneous redox reactions and their application in electrochemical cells. Refer to the textbook for information about half-reactions and electrode potentials. A table of standard reduction potentials is found in the textbook. The overall voltage of a cell may be enhanced by linking voltaic cells into a series circuit with the cathode of one cell attached with a wire to the anode of another cell. Because the metal strips can be reused, do not include them in your disposal costs.

## MATERIALS for Rescue Adjuncts, Inc.

### REQUIRED ITEMS
*(You must include all of these in your budget.)*

| | |
|---|---|
| Lab space/fume hood/utilities | 15 000/day |
| Standard liquid disposal fee | 20 000/L |

### REAGENTS AND ADDITIONAL EQUIPMENT
*(Include in your budget only what you'll need.)*

| | |
|---|---|
| 0.5 M Al₂(SO₄)₃ | 100/mL |
| 0.5 M CuSO₄ | 100/mL |
| 0.5 M MgSO₄ | 100/mL |
| 0.5 M SnCl₂ | 100/mL |
| 0.5 M ZnSO₄ | 100/mL |
| Metal strips: Al, Cu, Mg, Sn, Zn | 1 000/each |
| 150 mL beaker | 1 000 |
| 250 mL beaker | 1 000 |
| 100 mL graduated cylinder | 1 000 |
| Emery cloth | 1 000 |
| Filter paper | 500/piece |
| Glass stirring rod | 1 000 |
| Porous cup | 2 000 |
| Ring stand with buret clamp | 2 000 |
| Salt bridge | 2 000 |
| Tongs or forceps | 500 |
| Voltage probe | 3 000 |
| Voltmeter | 3 000 |
| Wire with alligator clips | 500 |

\* No refunds on returned chemicals or unused equipment.

### FINES

| | |
|---|---|
| OSHA safety violation | 2 000/incident |